Routledge Guides to the Great Books

The Routledge Guidebook to Galileo's *Dialogue*

The publication in 1632 of Galileo's *Dialogue on the Two Chief World Systems, Ptolemaic and Copernican* marked a crucial moment in the scientific revolution and helped Galileo become the father of modern science. The *Dialogue* contains Galileo's mature synthesis of astronomy, physics, and methodology, and a critical confirmation of Copernicus's hypothesis of the earth's motion. However, the book also led Galileo to stand trial with the Inquisition, in what became known as the greatest scandal in Christendom.

In *The Routledge Guidebook to Galileo's Dialogue*, Maurice A. Finocchiaro introduces and analyzes:

- The intellectual background and historical context of the Copernican controversy and Inquisition trial;
- The key arguments and critiques that Galileo presents on both sides of the "dialogue";
- The *Dialogue*'s content and significance from three special points of view: science, methodology, and rhetoric;
- The enduring legacy of the *Dialogue* and the ongoing application of its approach to other areas.

This is an essential introduction for all students of science, philosophy, history, and religion wanting a useful guide to Galileo's great classic.

Maurice A. Finocchiaro is Distinguished Professor of Philosophy, Emeritus; University of Nevada, Las Vegas. He has written and translated numerous works on Galileo and the history of science including *Galileo on the World Systems: A New Abridged Translation and Guide* (1997) and *The Essential Galileo* (2008).

ROUTLEDGE GUIDES TO THE GREAT BOOKS

Series Editor: Anthony Gottlieb

The Routledge Guides to the Great Books provide ideal introductions to the work of the most brilliant thinkers of all time, from Aristotle to Marx and Newton to Wollstonecraft. At the core of each Guidebook is a detailed examination of the central ideas and arguments expounded in the great book. This is bookended by an opening discussion of the context within which the work was written and a closing look at the lasting significance of the text. *The Routledge Guides to the Great Books* therefore provide students everywhere with complete introductions to the most important, influential and innovative books of all time.

Available:

Aristotle's Nicomachean Ethics Gerard J. Hughes
Galileo's Dialogue Maurice A. Finocchiaro
Hegel's Phenomenology of Spirit Robert Stern
Heidegger's Being and Time Stephen Mulhall
Locke's Essay Concerning Human Understanding E. J Lowe
Mill's On Liberty Jonathan Riley
Plato's Republic Nickolas Pappas
Wittgenstein's Philosophical Investigations Marie McGinn
Wollstonecraft's A Vindication of the Rights of Woman Sandrine Bergès

Forthcoming:

De Beauvoir's The Second Sex Nancy Bauer
Descartes' Meditations on First Philosophy Gary Hatfield
Hobbes' Leviathan Glen Newey

Routledge Guides to the Great Books

The Routledge Guidebook to Galileo's *Dialogue*

Maurice A. Finocchiaro

Routledge
Taylor & Francis Group

LONDON AND NEW YORK

First published 2014
by Routledge
4 Park Square, Milton Park, Abingdon, Oxon OX14 4RN
605 Third Avenue, New York, NY 10017

Routledge is an imprint of the Taylor & Francis Group, an informa business

British Library Cataloguing in Publication Data
A catalogue record for this book is available from the British Library

Library of Congress Cataloging in Publication Data
Finocchiaro, Maurice A., 1942-
The Routledge guidebook to Galileo's Dialogue / Maurice A. Finocchiaro.
pages cm. -- (The Routledge guides to the great books)
Includes bibliographical references and index.
1. Galilei, Galileo, 1564-1642. Dialogo dei massimi sistemi. 2. Solar system.
3. Astronomy. I. Title.
QB41.G138F56 2013
520--dc23
2013005749

ISBN: 978-0-415-50367-9 (hbk)
ISBN: 978-0-415-50368-6 (pbk)
ISBN: 978-0-203-70030-3 (ebk)

Typeset in Garamond
by Taylor and Francis Books

CONTENTS

ACKNOWLEDGMENTS

This book has been in the making for about four decades, and traces of it can be found in all my previous works on Galileo. Thus, my debts to scholars and institutions are massive, and cannot all be mentioned here. However, allow me to refer readers to the prefaces to my previous books, and to mention explicitly the more important cases and add the more recent ones. For long-standing encouragement and support, I thank John Heilbron, William Wallace, and Robert Westman. For valuable interactions and exchanges, my gratitude goes to Antonio Beltrán Marí, Mario Biagioli, Richard Blackwell, Massimo Bucciantini, Michele Camerota, Maurice Clavelin, George Coyne, Annibale Fantoli, Andrea Frova, Owen Gingerich, David Hill, Nicholas Jardine, Michel-Pierre Lerner, W. Roy Laird, Franco Motta, Ron Naylor, Paolo Palmieri, Mauro Pesce, Jürgen Renn, Arcangelo Rossi, Michael Segre, Peter Slezak, and Albert van Helden. For specific feedback on the present project, my appreciation goes to Robert Hankinson and Michael Shank. For commenting on the entire manuscript of this book, I am indebted to Albert DiCanzio. For editorial advice and substantive comments, I am grateful to Routledge's own editors, Andy Humphries and Anthony Gottlieb. For moral and material support at my university, I am grateful to my colleagues David Beisecker, Ian Dove, David Forman, Todd Jones, William Ramsey, Paul Schollmeier, and James Woodbridge.

Acknowledgments also go to the following institutions. For financial support in the form of research grants and fellowships, my gratitude goes to the National Science Foundation, National Endowment for the Humanities, and the Guggenheim Foundation. Special recognition goes to the University of Nevada, Las Vegas for several sabbatical and research leaves while I was teaching there, and for continued office privileges after my formal retirement from teaching. For sponsorship of lectures on Galileo, my appreciation goes to the following universities: Barcelona, Bologna, California-Berkeley, California-Los Angeles (UCLA), California-San Diego, Cambridge, Columbia, Florence, Harvard, London, Melbourne, Mexico City (UNAM), Minnesota, New South Wales, Oxford, Padua, Pittsburgh, Rome-La Sapienza, Stanford, Sydney, and Tel-Aviv.

Furthermore, I thank the publishers of my Galileo books from which I have borrowed freely, but in a general sort of way: the University of California Press, which published *The Galileo Affair*, *Galileo on the World Systems*, and *Retrying Galileo 1633–1992*; Springer Science and Business Media, publisher of *Galileo and the Art of Reasoning* and *Defending Copernicus and Galileo*; and Hackett Publishing Company, publisher of *The Essential Galileo*. Similarly, I have benefited from previous editions and translations of Galileo's *Dialogue*, often adapting without acknowledgment the information they provide; but where my debt is more direct, I explicitly give appropriate references to them (for which see the Abbreviations list): Webbe (1635), Salusbury (1661), Strauss (1891), Favaro (1890–1909), Santillana (1953), Pagnini (1964), Drake (1967, 2001), Sosio (1970), Sexl and von Meyenn (1982), Fréreux and de Gandt (1992), Beltrán (1994, 2003), Besomi and Helbing (1998), and Shea and Davie (2012).

Last but not least, specific formal acknowledgments are due to some publishers and copyright holders. For quotations of many Galilean passages, and for adaptations of many other parts, I acknowledge the permission by the University of California Press to use such material from: Galileo Galilei, *Galileo on the World Systems: A New Abridged Translation and Guide*, ed. and trans. by

M.A. Finocchiaro, © 1997 by the Regents of the University of California, published by the University of California Press. For the four diagrams in the discussion of the paths of sunspots (chapter 6, section IIIC3), I acknowledge the permission by the University of Chicago Press to reproduce them from A.M. Smith, "Galileo' Proof for the Earth's Motion from the Movement of Sunspots," *Isis*, 76 (1985): 543–51, © 1985 by the History of Science Society, published by the University of Chicago Press.

SERIES EDITOR'S PREFACE

"The past is a foreign country," wrote British novelist, L. P. Hartley: "they do things differently there."

The greatest books in the canon of the humanities and sciences can be foreign territory, too. This series of guidebooks is a set of excursions written by expert guides who know how to make such places become more familiar.

All the books covered in this series, however long ago they were written, have much to say to us now, or help to explain the ways in which we have come to think about the world. Each volume is designed not only to describe a set of ideas, and how they developed, but also to evaluate them. This requires what one might call a bifocal approach. To engage fully with an author, one has to pretend that he or she is speaking to us; but to understand a text's meaning, it is often necessary to remember its original audience, too. It is all too easy to mistake the intentions of an old argument by treating it as a contemporary one.

The *Routledge Guides to the Great Books* are aimed at students in the broadest sense, not only those engaged in formal study. The intended audience of the series is all those who want to understand the books that have had the largest effects.

AJG
October 2012

AUTHOR'S PREFACE

When Routledge approached me about writing this book, at first I could not help wondering whether there was anything new for me to write. For I had already published several books dealing implicitly or explicitly with Galileo's *Dialogue on the Two Chief World Systems, Ptolemaic and Copernican*. For example, *Galileo and the Art of Reasoning* (1980) is an analytical commentary to the *Dialogue*, which uses the material therein to elaborate a case study in logical theory and philosophy of science. *The Galileo Affair* (1989) is a collection of the essential documents about the trial of Galileo, which provides some crucial historical context for his book. *Galileo on the World Systems* (1997) is an abridged translation of about half of the *Dialogue*, together with introductions, appendices, and annotations, aiming to guide the reader through the text. And *Retrying Galileo, 1633–1992* (2005) is a survey of the facts, documents, and issues of the controversy about Galileo's trial, which is also about the book—the *Dialogue*—that caused the trial and was banned as a result.

However, I quickly realized that none of these books provided what was called for: a reconstruction of the whole main argument in Galileo's *Dialogue*. Such a reconstruction makes up the heart of the present book (Chapters 4–7). It is a summary of Galileo's main argument that aims to be complete, self-contained, and detailed, and to provide interpretation, evaluation, and analysis. It is sufficiently faithful to the original to retain the *Dialogue*'s basic structure and topical progression, but also sufficiently critical (constructively critical) to improve clarity whenever possible, by elucidating obscurities, resolving unhelpful ambiguities, avoiding needless confusions, and making explicit the hidden assumptions and presupposed information. It is also intended to be sufficiently detailed not to miss anything essential, but sufficiently streamlined to avoid digressions (either Galileo's own, or those of scholars seeking to delve deeper).

These features of the reconstruction are reflected in the numbering system which I use to label the various sections of Chapters 4–7. Rather than following the usual style of using these chapter numbers followed by decimals, I use the Roman numerals I, II, III, and IV, to correspond, respectively, to the four "Days" into which the *Dialogue* is divided. Then, within each such "Day," the Roman numerals are followed by capital letters A, B, C, and/or D to designate the main parts into which it can be further subdivided, making explicit a subdivision which is merely implicit in the Galilean text. Finally, the capital letters are followed by a third set of symbols, Arabic numerals, to render explicit some finer thematic distinctions, which Galileo does not make explicit, but which are more or less obvious to the careful reader. Thus, for example, the passage discussing the "extruding power of whirling" is labeled "IIB11," to indicate that it discusses the eleventh of the traditional arguments against the earth's daily axial rotation, which arguments comprise strand "B" of "Day II." These tripartite labels are meant to guide readers not only by giving them a sense of structure and progression as they read through Chapters 4–7, but also to enable them to immediately identify and contextualize various Galilean passages as these are discussed or mentioned in the other chapters of the present work. Finally, note that in chapter 6 there is some switching back and forth of the "B" and "C" strands, to reflect Galileo's own non-uniform order of exposition in Day III.

This reconstruction can be read independently of the rest of the present book, by anyone who has had occasion to read, reflect upon, or otherwise be introduced to the *Dialogue*. For those who have not been so introduced, or who lack the background preparation to read Galileo's book or my reconstruction, Chapters 1–3 provide the main preliminaries. Note that I really begin with fundamentals, such as why one should want to read the *Dialogue* in the first place, and why it is a great book (Chapter 1). The intellectual background (Chapter 2) should be no news to experts, but will be essential for the beginner. On the other hand, the account of the historical context (Chapter 3) is addressed to both beginners and specialists, insofar as it is meant to be comprehensible and useful to the former, as well as original and well documented for the latter. Similarly, Chapters 8–10 are meant for both novices and experts. They delve more explicitly, systematically, deeply, and critically into three special aspects of the *Dialogue* that are barely touched upon in the reconstruction; they analyze the book's content and significance from the points of view of scientific achievement, methodology or epistemology, and the arts of rhetoric. The conclusion (Chapter 11) too is for both laypersons and scholars; it discusses long-range historical repercussions and enduring cultural lessons of interest to both.

Finally, the notes and references require a few words of explanation. The endnotes are reserved for discursive notes and references that are complex and/or multiple. On the other hand, simple references to various passages of the *Dialogue* are usually given parenthetically within my exposition. For this purpose, I use primarily Stillman Drake's English translation in the 2001 Modern Library edition (Galilei 2001, abbreviated DML), rather than the 1967 University of California edition (Galilei 1967, abbreviated DCA). Frequently, quotations of Galilean passages are taken from my own abridged translation (Galilei 1997, abbreviated FIN), or made anew directly from the Italian text of the critical edition in Galileo's collected works edited by Favaro (Galilei 1897, abbreviated FAV); and, occasionally, passages are quoted from Santillana (Galilei 1953b) or Shea-Davie (Galilei 2012); in such cases, references are obviously also given to them. For simplicity's sake, passages from the *Dialogue* are usually referenced with at most two citations, one from DML, plus another from the source of the quotation. However, as a service to readers, complete cross-references to all relevant passages in DML, DCA, FAV, and FIN are provided in a table in the Appendix, which readers should consult for further explanations. Similarly, they should consult the list of Abbreviations.

ABBREVIATIONS

References to various editions and translations of Galileo's *Dialogue* are so numerous in this book that I shall abbreviate them by using the name of the editor(s) or translators(s) followed immediately by the page number(s). This will also display proper credit to these scholars. As customary, all these books are listed in the bibliography under Galileo's own name, with the entries being ordered by year of publication, and so they could also be referred to by using "Galilei" followed by a date. Although these scholars' names could be found by sifting through that list, to facilitate the process, here I provide a list of the correspondence between these scholars' names and those bibliographical entries. References to Favaro will sometimes have decimal numbers to designate line numbers. Note also that the Favaro references are further abbreviated FAV. Similarly, references to Drake's translation are also abbreviated further: DML for the Modern Library edition, and DCA for the University of California Press edition. And references to my abridged translation are further abbreviated FIN. Finally, note also that the Appendix gives complete cross-references for these four editions (FAV, DCA, DML, and FIN).

Beltrán   Galilei 1994
Beltrán-IT   Galilei 2003

Besomi-Helbing	Galilei 1998, vol. 2
DCA (short for Drake-California)	Galilei 1967
DML (short for Drake-Modern-Library)	Galilei 2001
FAV (short for Favaro)	Galilei 1897 (= Galilei 1890–1909, vol. 7)
FIN (short for Finocchiaro)	Galilei 1997
Fréreux-Gandt	Galilei 1992
Pagnini	Galilei 1964, vols. 2–3
Salusbury	Galilei 1661
Santillana	Galilei 1953b
Sexl-Meyenn	Galilei 1982
Shea-Davie	Galilei 2012
Sosio	Galilei 1970
Strauss	Galilei 1891
Webbe	Galilei 1635

Part I

PRELIMINARIES TO READING
THE *DIALOGUE*

1

GENERAL RELEVANCE

From prehistoric times until the beginning of the sixteenth century, almost everybody believed that the earth stood still at the center of the universe and all heavenly bodies revolved around it. By the end of the seventeenth century, most astronomers, physicists, and philosophers had come to believe that the earth was the third planet circling the sun once a year and spinning around its own axis once a day. Nowadays, after more than three centuries of accumulating evidence, this modern geokinetic view is known to be true beyond any reasonable doubt. However, the earlier geostatic view had been a very reasonable belief; for two millennia the earth's motion had been rejected as untenable. Then, for about two centuries the discussion of the relative merits of the two views was the subject of heated debate. In fact, the transition was a slow, difficult, and controversial process. We may fix its beginning with the publication in 1543 of Nicolaus Copernicus's book *On the Revolutions of the Heavenly Spheres*, and its completion with the publication in 1687 of Isaac Newton's *Mathematical Principles of Natural Philosophy*.

This discovery of the motion and noncentral location of the earth involved not only a key astronomical, physical, and cosmological fact, but also the discovery of the most basic laws of nature, such as the physical laws of inertia, force and acceleration, action and reaction, and universal gravitation. This discovery was also interwoven with the clarification of some key principles of scientific method. It represents, therefore, without a doubt, the most significant breakthrough in the whole history of science; accordingly, the series of developments that started with Copernicus in 1543 and ended with Newton in 1687 may be labeled the scientific revolution.[1]

More generally, it would be no exaggeration to say that this transition represents the most important intellectual transformation in human history. One reason for this involves the world-wide repercussions of the scientific revolution itself: science is the only cultural force that has managed to dominate human societies and cultures in all parts of the earth. Another reason involves the interdisciplinary character of the transition from a geostatic to a geokinetic world view: the transformation affected not only many branches of science, but had profound effects on disciplines and activities other than science, such as philosophy, theology, religion, art, literature, technology, industry, and commerce; indeed it changed mankind's self-image in general. We may thus also call this transition the Copernican revolution, if we want a label that leaves open its broad ramifications outside science and gives due credit to the one thinker whose contribution initiated the process.[2]

Galileo Galilei (1564–1642) was one of the protagonists of this historical development.[3] Thus, he is regarded as one of the founders of modern science, along with Copernicus (1473–1543), Johannes Kepler (1571–1630), René Descartes (1596–1650), Christiaan Huygens (1629–95), and Isaac Newton (1642–1727). Indeed, Galileo is often singled out as the most pivotal of these founders and called the Father of Modern Science. Although many people have repeated this characterization, it is important that it originates in the judgment of practicing scientists themselves, such as Albert Einstein (1954: 271) and Stephen Hawking (1988: 179, 2002: xvii).

In physics, Galileo pioneered the experimental investigation of motion; he formulated, clarified, and systematized many of the basic concepts and principles needed for the theoretical analysis of motion; and he discovered the laws of falling bodies, including free fall, descent on inclined planes, and pendulums. In astronomy, he introduced the telescope as an instrument for systematic observation; he made a number of crucial observational discoveries, such as mountains of the moon, satellites of Jupiter, phases of Venus, and sunspots; and he understood the cosmological significance of these observational facts and gave essentially correct interpretations of many of them. He was also an inventor, making significant contributions to the devising and improvement of such instruments as the telescope, microscope, thermometer, and pendulum clock.

Regarding the scientific method, he pioneered several important practices, such as the use of artificial instruments (like the telescope) to learn new facts about the world, and the active intervention into and manipulation of physical phenomena in order to gain access to aspects of nature that are not detectable without such experimentation. He also contributed to the establishment and extension of other more traditional methodological practices, such as the use of a mathematical approach in the study of motion. And he contributed to the explicit formulation and clarification of important methodological principles, such as the setting aside of biblical assertions and religious authority in scientific inquiry.

Galileo's *Dialogue on the Two Chief World Systems, Ptolemaic and Copernican*, first published in 1632, is one of the most important texts of the Copernican and scientific revolutions. It constitutes his mature synthesis of astronomy, physics, and scientific methodology. It also provided a significant confirmation of the Copernican theory. This confirmation required a two-fold approach: defending the geokinetic theory from the many objections based on Aristotelian physics, naked-eye astronomical observation, and traditional epistemology; and supporting the earth's motion with reasons stemming from his new physics, from telescopic observation, and from the methodological analysis of contextualized epistemological problems.

The *Dialogue* is also the work that triggered the trial of Galileo by the Roman Catholic Inquisition in 1633, ending with his condemnation as a heretic and the banning of the book. Publication of the book had been problematic in light of a series of events that began in 1613 and included a related series of Inquisition proceedings in 1615–16. This twenty-year sequence of developments may be called the Galileo affair, and so the book is also an important document in this tragic but instructive episode. The details of this historical context will be examined later (Chapter 3).

Moreover, the 1633 condemnation of Galileo gave rise to a new controversy that continues to our own day (see Finocchiaro 2005b, McMullin 2005a). This cause célèbre is a controversy about Galileo's trial—its facts, causes, consequences, and lessons. This subsequent controversy partly reflects the issues of the original one, but primarily involves additional issues that have had a formative influence on modern Western culture: whether, how, and why the condemnation of Galileo was right or wrong; whether or not it proves the incompatibility between science and religion; and what can be learned from it regarding the interaction between individual freedom and institutional authority.

This subsequent cause célèbre may also be called the Galileo affair. Thus, to be terminologically clear, the latter phrase may refer to either the original episode spanning the years 1613–33, or the subsequent and ongoing controversy, or both. But the substantive point is that the *Dialogue* is also an important document for the ongoing cause célèbre about the original trial.

The introductory remarks so far are meant to summarize and highlight the book's perennial and universal relevance by way of its connections with events such as the scientific revolution, the Copernican revolution, and the Galileo affair (in both senses). However, the book's general relevance can be elaborated in another way that connects it to a number of mental activities or intellectual practices of fundamental significance.

The most ubiquitous feature of the *Dialogue* is critical reasoning, taking this term to mean reasoning aimed at the interpretation, analysis, evaluation, or self-reflective presentation of arguments. Reasoning, in general, is the mental process of interrelating thoughts in such a way that the truth of some depends on the truth of others, by the former being based on or following from the latter. And an argument may be defined as a piece of reasoning aimed to justify a conclusion by supporting it with reasons, or defending it from objections, or both.[4]

Galileo was writing at a time when the new Copernican world view was not yet conclusively established beyond any reasonable doubt, but was in a context where the situation was still dynamic and controversial. Thus, to form an intelligent opinion on the topic required more than merely observation, experiment, calculation, or deduction; what was required was reasoning, judgment, analysis, evaluation, and argumentation. So it is not surprising that he felt the most fruitful thing to do was to undertake a critical examination of the arguments on both sides of the controversy.

Galileo's main conclusion is clearly that the earth's motion is more probable, or more likely to be true, than the earth's rest. To arrive at this conclusion, his approach is to show that the pro-Copernican arguments are stronger than the anti-Copernican ones; that is, that the reasons for believing the earth to be in motion are better than those for believing it to stand still; or again, that the evidence supporting the geokinetic idea outweighs that supporting the geostatic one. Galileo's conclusion does *not* mean that Copernicanism is either clearly true, or certainly true, or absolutely true, or demonstrably true; nor does it mean that there are no reasons for believing the earth to stand still; nor that the geostatic arguments are worthless.

In other words, the book's key thesis may be formulated as one about the relative merits of the arguments on each side: that the geokinetic arguments are stronger than the geostatic ones. Moreover, this thesis is substantiated and not merely asserted; and the substantiation proceeds by the self-reflective presentation, interpretation, analysis, and evaluation of the respective arguments. Thus, the book's key thesis could also be formulated as saying that the critical reasoning of the Copernicans is stronger that that of the Ptolemaics. In short, critical reasoning is a key part of the book's *content*, as well as of the book's *approach*.

These two formulations of the book's main thesis correspond to each other, but reflect different (not opposite or incompatible) points of view. That is, taking the point of view of the natural phenomena, the book is claiming that the earth's motion is more likely to be true than the earth's rest. Taking the point of view of critical reasoning, the book is claiming that the geokinetic arguments are stronger than the geostatic ones. These formulations also correspond to my earlier claims that the book provided a significant confirmation of the geokinetic hypothesis, and that it represented a synthesis of the new physics, astronomy, and methodology. All these interpretations (in terms of interdisciplinary synthesis, Copernican confirmation, critical reasoning, and probability of the earth's motion) will be elaborated, defended, and interrelated later (Chapter 8).

Despite the prevalence of critical reasoning, it is equally apparent that Galileo's *Dialogue* is full of methodological or epistemological reflections. By methodological or epistemological reflections (which will be regarded as synonymous here), I mean discussions meant to clarify, formulate, and apply general problems and principles about the nature of truth and knowledge and about the proper procedure to follow in the search for truth and in the acquisition of knowledge; by calling them reflections I mean to convey the idea that these discussions arise in the context of some particular investigation about truth in, or knowledge of, physical reality, and so they have the function of helping one understand better what one is doing or deciding what one ought to be doing.[5]

In other words, problems and principles about truth, method, and knowledge are constantly discussed in the book, not because Galileo intends to write an abstract treatise about the nature of these concepts, but because the specific astronomical and physical issues are so fundamental that they frequently raise questions about how one is proceeding and what is the proper way to proceed in scientific inquiry. For example, there are discussions about the nature and proper role of authority, sensory observation and intellectual speculation, the limitations of human understanding, independent-mindedness and open-mindedness, simplicity, probability, experimentation, mathematics, artificial instruments, the Bible, divine purpose and human interest, and causal explanation. The methodological aspect of the *Dialogue* will be elaborated explicitly later (Chapter 9).

Galileo's *Dialogue* is also a goldmine of rhetoric, but here one must be especially careful. In this context, I take rhetoric to mean the theory and practice of verbal communication, involving such forms and techniques as persuasive argumentation, eloquent expression, beautiful language, imaginative portrayal, emotional description, bare assertion, nuanced assertion, repetition, double entendre, wit, satire, humor, and ridicule. Notice that I am *not* equating rhetoric with the art of deception in general, and the skill of making the weaker argument appear stronger in particular; so understood, rhetoric would be an inherently objectionable activity, whereas my definition allows for both good and bad rhetoric.

The wealth of the book's rhetoric derives in part from its universalist aim, which implies that Galileo is addressing several audiences at once. It derives in part from its dialogue form, which means that there is a certain amount of drama unfolding before the reader. It also stems from the controversial character of the scientific and epistemological issues discussed, which means that we are witnessing a brilliant polemic. It also stems from the context of Galileo's struggle with the Catholic Church; this means that in writing the book he was taking considerable risks and could not always safely say what he meant or mean what he said. It originates, to some extent, from the fact that the practice of science at that time was socially and financially dependent for the most part on the patronage of princes; this means, generally, that Galileo's career was partly that of a courtier, and specifically that his book represented an action in an intricate network of patronage involving the Tuscan Medici court in Florence and the Vatican court of Pope Urban VIII in Rome. Finally, the rhetoric originates to some extent from the fact that Galileo was a gifted writer who poured his heart and soul into this work, so much so that many passages achieve a high degree of literary and aesthetic value. The rhetorical aspect of the *Dialogue* will be examined explicitly later (Chapter 10).

In summary, Galileo's *Dialogue* can and should be read for what it tells us about the history of the Copernican revolution, the scientific revolution, the trial of Galileo, and the ongoing Galileo affair, and also for what it can teach us in general about critical reasoning, scientific methodology, and the arts of rhetoric.

NOTES

1 For valuable accounts centered around this theme, see Hall 1954, Lindberg and Westman 1990, H.F. Cohen 1994, and Henry 1997. Talk of "revolution" can be misleading, as Westman (2011: 22) has recently reminded us; this is especially true if one approaches the topic from the viewpoint of political revolutions, or if

one assumes a particular conception of scientific revolutions. The best known example of the latter is Kuhn's (1962, 1970) account, which views scientific revolutions as transitions from one previously established "paradigm" to a new, incommensurable, and incompatible paradigm. However, such problems can be avoided by avoiding the political analogy and the Kuhnian conception.

2 For a classic example of this type of account, see Kuhn 1957. Like talk of the scientific revolution, talk of the Copernican revolution can be misleading, as Westman 2011 has recently stressed. However, the notion of a Copernican revolution is much more firmly established in all kinds of contexts. At any rate, as for the case of the scientific revolution, problems can also be avoided by avoiding political and Kuhnian presuppositions.

3 For good scholarly general accounts, see Geymonat 1965, Drake (1978, 1990), Heilbron 2010; for interesting and basically accurate popular accounts, see Seeger 1966, Ronan 1974, Sobel 1999.

4 For more details on these concepts, see Scriven 1976, Fisher and Scriven 1997, Finocchiaro (2005a: 92–108, 292–326).

5 What I am calling methodological or epistemological reflections correspond to what in the literature on thinking are variously labeled metacognition, reflective thinking, and applied epistemology; see, e.g., Finocchiaro (1980: 103–79, 1997: 335–39, 2005a: 92–108).

2

INTELLECTUAL BACKGROUND

The world view generally accepted until the Copernican revolution involved many details to be explained shortly, but contained two fundamental theses that generated referring labels. One thesis claimed that the earth is motionless, and so we may speak of the *geostatic* view or, more simply, geostaticism.[1] The other claim asserted that the earth is located at the center of the universe, and so we may call it the *geocentric* theory or, more simply, geocentrism.

Although we know now that the geocentric view is not true, it corresponds, even today, to everyday observation and common sense intuition; and, although it has this natural appeal, its technical elaboration was the result of arduous work by some of the greatest thinkers of antiquity. In particular, two individuals made such important contributions that their names became synonymous with this view. Aristotle (384–322 B.C.), a pupil of Plato, lived in Athens during the period of classical Greek civilization; he contributed primarily by elaborating the cosmology, physics, general philosophical principles, and qualitative astronomical ideas of the geostatic view. Ptolemy lived in Alexandria in the second century A.D., at the end of the Hellenistic phase of Greek culture; he contributed primarily the mathematics and quantitative details of the astronomical system, forging a synthesis of the observational, mathematical, and theoretical discoveries of the five intervening centuries. Thus, we may also label the old view the *Aristotelian* or *Ptolemaic* theory. Furthermore, since the Aristotelians acquired the nickname of Peripatetics, geocentrism was also traditionally labeled the *Peripatetic* world view; and it is sometimes so labeled in the *Dialogue*.

Obviously, the geocentric view was not a simple and monolithic entity, but rather was a theory that underwent two thousand years of explicit historical development. The version expounded below is not a synopsis of any one treatise, but rather a reconstruction of the most widely held beliefs at the beginning of the sixteenth century, in a form useful for understanding and appreciating Galileo's *Dialogue*.[2]

2.1 COSMOLOGY

Let us begin with the earth's *shape*. The geostatic view held that the earth is a sphere, and so its surface is not flat but round; this is, of course, true. In fact, the arguments proving this fact were already known to Aristotle, and can be found in his writings. Although uneducated persons or primitive peoples at the time of Aristotle or Galileo may have believed that the earth is flat, scientists and philosophers had settled the question much earlier; thus, it should be clear that the Copernican controversy had nothing to do with the shape of the earth, but with its location and behavior.

The shape which became part of that controversy was the shape of the whole universe. In fact, the old view held that the universe was a sphere much larger than the earth, but of *finite* size, the size being only slightly larger than the orbit of the outermost planet. The stars were all at the same distance from the center, being attached to the surface of the so-called *stellar* or *celestial* sphere; this stellar sphere enclosed the whole universe, and outside this sphere there was nothing physical. This contrasts with the modern view that the universe is infinite, space goes on without end, and stars are scattered everywhere in infinite space; thus, it does not even make sense to speak of the shape or center of the universe.

Nevertheless, the finite, spherical universe was based on the same set of observations which led to the belief that at the center of the stellar sphere was located the motionless earth. This was the phenomenon of apparent *diurnal motion*: the earth feels to be at rest; the whole universe appears to move daily around the earth in a westward direction; thousands of stars visible with the naked eye at night appear to undergo no change in size or brightness, but seem to be at a fixed distance from us; they appear to move in unison, so that their relative positions remain fixed; they appear to move in circles which are larger for stars lying closer to the equator and smaller for those lying closer to the poles; in short, the stars appear to move as if they were attached to a sphere which was rotating daily westward around a motionless earth at the center. Given the plausible principle that what appears to normal observation corresponds to reality, one had the basic argument supporting the basic tenets of the geostatic world view.

In the finite, spherical universe, position or location or place had an absolute meaning. The geometrical center of the stellar sphere was a definite and unique place, and so was its surface or circumference; and between the center and the circumference, various layers or spherical shells defined various intermediate positions. The part of the universe outside the earth was called *heaven* in general, and to distinguish one heavenly region from another they spoke of different heavens (in the plural). For example, the stellar sphere was the highest heaven, which meant the most distant one from the earth, and was also called the *firmament*; whereas the closest heaven was the spherical layer to which the nearest heavenly body (the moon) was attached, and so the lunar sphere was the first heaven. Between the lunar and the stellar spheres, six other particular heavens or heavenly spheres were distinguished; one was for the sun, while there was one for each of the other five known planets (Mercury, Venus, Mars, Jupiter, and Saturn). More details about the motion of the planets will be discussed later.

Here, it is important to distinguish a *heavenly sphere* from a *heavenly body*: a heavenly sphere was one of the eight nested spherical layers surrounding the central earth, each of which was the region occupied by a particular heavenly body or group of heavenly bodies, and to each of which these heavenly bodies were respectively attached; whereas a heavenly body was a term referring to either the sun, moon, a planet, or one of the thousands of fixed stars. The heavenly bodies were considered to be spherical in shape, and were thus spheres in their own right; but the term "heavenly spheres" referred only to the spheres concentric with the center of the universe to which the "heavenly bodies" were attached.

The terrestrial region, too, had its own layered structure. This is related to a three-fold meaning for the term *earth*. In saying earlier that the earth is a sphere, I was referring to the terrestrial globe consisting of land and oceans. This globe is a sphere, not in the sense of a perfect sphere, but only approximately, because the land is above the water and is full of mountains and valleys. Such an approximation is, of course, very good because the height of even the tallest mountain is insignificant compared to the earth's radius. However, it was only natural to distinguish water from earth, taking the latter term in the sense of just land, rocks, sand, and minerals; when so understood, "earth" is obviously only a part of the whole globe. Next, it was also natural to count the air or atmosphere surrounding the globe as part of the terrestrial region; and so by earth one could also mean the whole region of the universe near the terrestrial globe, up to but excluding the moon. In short, the term earth had three increasingly broader meanings: it could refer to just the solid part of the terrestrial globe; or it could refer to the whole globe consisting of both land and oceans; or it could refer to the whole terrestrial region of land plus oceans plus atmosphere.

Terminology aside, the substantive point is that the earth (namely, the place where mankind lives) is not a body of uniform composition, but has three main parts; it contains a solid part, a liquid part, and a gaseous part. These three parts (namely, *earth*, *water*, and *air*) were labeled *elements* to signify their fundamental importance. In regard to the arrangement of these terrestrial substances, the element earth sinks in water, and so earth must extend to the central inner core of the world and must make up most of what exists below the surface; on the other hand, most of the surface of the globe is covered with water, and the element water mostly surrounds the element earth. This was expressed theoretically by claiming that the natural place of the element earth was a sphere immediately surrounding the center of the universe, and that the natural place of the element water was a spherical layer surrounding the innermost sphere. In the case of the air, simple observation tells us that it surrounds the spheres of the first two elements, and so its natural place was a third sphere surrounding the first two.

There was a fourth terrestrial element, to which the name *fire* was given; but it required a more roundabout explanation. Just as we see earth sink in water, and water fall down (as rain) through air, we see flames shoot upwards through air when something is burning; we also see currents of heat move upwards through air during hot summer days, and smoke generally rise; and we see trapped fire escape upwards in volcanic eruptions. Such observations were taken as evidence that the natural place of fire was a fourth spherical layer above the atmosphere and just below the lunar sphere.

The existence of the element fire was also derived from some considerations about basic physical qualities. There were two fundamental pairs of physical opposites: hot and cold, and humid and dry. The element earth was a combination of cold and dry; the element water was a combination of cold and humid; and the element air was a combination of hot and humid. So there had to be a combination of hot and dry, and that was what constituted the element fire.

Like position, direction had an absolute meaning in the finite universe. There were three basic directions: toward the center of the universe, called *downward*; away from the center, called *upward*; and around the center. Thus, one important way of classifying motions was in these cosmological terms: bodies could and did move toward, away from, and around the center of the universe.

Geometrically speaking, motion could be simple or mixed. Simple motion was motion along a simple line. A simple line was defined as a line every part of which is congruent with any other part. Thus, there were supposedly only two such lines, circles and straight lines; and there were two types of simple motion, straight and circular motion. Mixed motion was motion which is neither straight nor circular.

Another way of classifying motions was in terms of the motions characteristic of the various elements, namely the motions which the elements undergo spontaneously. This was meant to correspond with the two other classifications in the following manner. For example, earth and water characteristically moved straight downwards, while air and fire characteristically moved straight upwards. Now, since heavenly spheres and heavenly bodies moved characteristically with circular motion around the center, this meant that they were composed of a fifth element; the term *aether* or *quintessence* was used to refer to this heavenly element.

Finally, another important classification was in terms of the opposition between natural and violent motions. *Violent motion* was motion caused by some external action; *natural motion* was motion which a body underwent because of its nature, so that the cause was internal. For example, the downward motion of earth and water, the upward motion of air and fire, and the circular motion of heavenly spheres and heavenly bodies were all cases of natural motion; on the other hand, rocks thrown upwards, rain blown sideways by the wind, a cart pulled by a horse, and a ship sailing over the sea were all cases of violent motion.

More fundamentally, motion was the opposite of rest. Rest was the natural state of bodies, and so all motion presupposed a force in some way. Natural motion was essentially the motion of a body toward or within its proper place; only when displaced from its proper place by some force would a terrestrial body engage in natural motion up or down; and only if started by some mover would a heavenly sphere rotate around the center of the universe, thus carrying its planet or stars in circular motion. On the other hand, violent motion was motion which was not toward the body's proper place, and such motion could only happen by the constant operation of a force.

From what has already been said, it follows that earth and heaven were very different; this radical difference was enshrined in an idea that needs to be made explicit and deserves a special label. The key term is the *earth–heaven dichotomy*; but one could equivalently speak of the dichotomy between the earthly or terrestrial or sublunary or elemental region of the universe on the one hand, and the heavenly or celestial or superlunary or aethereal region on the other.

We have already seen that one difference between the two regions was location, which was absolute in the finite, spherical universe; terrestrial bodies occupied the central region below the moon, heavenly bodies occupied the outer region from the lunar to the stellar sphere. Similarly, we have also seen that there was another difference in regard to natural motions; earthly bodies moved naturally straight toward or away from the center of the universe, whereas celestial bodies moved circularly around the same center. We have also seen that the two regions differed in regard to the elements of which bodies were composed. Sublunary bodies were made of earth, water, air, or fire, or a mixture thereof. On the other hand, in the superlunary region things were made of aether, or various concentrations thereof; that is, aether in low concentration made up the heavenly spheres, which were actually invisible; whereas aether in a highly concentrated state generated the heavenly bodies we actually saw (moon, sun, planets, and stars).

Now, just as the natural places and natural motions of the two regions obviously corresponded to each other, the elements in the two regions also corresponded to the natural places and motions. That is, the natural places and motions of terrestrial bodies could be conceived as the essential properties of the terrestrial elements, while the natural places and motions of celestial bodies could be conceived as the essential properties of aether.

Moreover, other differences between earth and heaven could be defined in terms of additional properties of the different elements. For example, whereas superlunary substances had no weight, sublunary bodies obviously did. Or to be more exact, whereas aether was weightless, the sublunary bodies subdivided into two classes: earth and water had weight or *gravity*, and so were called *heavy bodies*; but air and fire had a tendency to go up or *levity*, and so were called *light bodies*. Moreover, aether was regarded as intrinsically luminous, namely capable of giving off its own light; but earthly elements were dark, namely incapable of emitting their own light. Even fire did not emit an inherent light of its own, but only temporarily produced light when in the process of escaping from lower regions to move to its natural place just below the lunar sphere.

Of the various differences between earth and heaven, two deserve special attention: natural motion, and susceptibility to qualitative change. To elaborate, one must first understand why the geostatic universe was *not* a trichotomy, given that after all there were three visible kinds of natural motions (downward for earth and water, upward for air and fire, and around the center for aether). The answer implied by what was stated earlier was that downward and upward natural motions were both straight, and so were conceived as two minor subspecies of the same fundamental kind—rectilinear motion. Geometrically, there were supposedly only two lines with the property that all parts are congruent with any other part, and they were the circle and the straight line; thus, what was common to both upward and downward natural motions (straightness) was more important than what distinguished them (toward and away from the universal center).

However, this geometrical reason was not the only justification for making the essential distinction to be two-fold (straight vs. circular natural motions), rather than three-fold (upward, downward, and around). There was also the cosmological reason that, unlike circular natural motion, straight natural motion could not be everlasting. For, once a rock had reached the center of the universe, its nature would make it remain there rather than continue moving past the center, which would constitute upward and thus unnatural motion for the rock. Similarly, once a fiery object had reached the region above the terrestrial atmosphere, just below the lunar sphere, it had reached its natural place and had no place to go; for to continue moving would bring it into the first heavenly sphere, which was reserved for the aethereal moon, and where the element fire could not subsist.

Finally, there was a theoretical reason why upward and downward natural motions could belong to the same fundamental region of the universe and were essentially different from natural circular motion. The theory in question was the theory of change and contrariety, according to which all change derives from contrariety, and no change can exist where there is no contrariety; by contrariety, this theory meant such oppositions as those between hot and cold and between dry and humid. Now, up and down, together with the related pair of light and heavy, was another fundamental contrariety.

It followed that a region full of bodies, some of which moved naturally downwards and some upwards, was bound to be full of all sorts of qualitative changes; and indeed observation obviously revealed that the terrestrial world is full of births, growth, decay, generation, destruction, weather and climatic changes, and so on. On the other hand, the circular, natural motion of the heavenly bodies was thought to have no contrary; moreover, the opposition between hot and cold and between dry and humid belonged only within the four terrestrial elements; thus the region of aether lacked any of the essential conditions for change. And observation confirmed that, too, because no physical or organic or chemical changes are easily detected in the heavens, and none were said to have ever been seen. The only essential phenomenon in the heavens was motion, but all heavenly motion was fundamentally regular: it involved the rotation of concentric spheres, which thus remained in place, so that there was not even change of place; what changed was only the relative position of the various bodies attached to the different celestial spheres.

2.2 PHYSICS

In the terrestrial region, the *natural state* of bodies was rest. To be more exact, it was rest at the proper place, depending on the elemental composition of the body: at the innermost core for the element earth; just above that for water; above water for air; and above air for fire. This meant that, whereas no cause was sought to explain why a body rested at its proper place, when a body was in motion or at rest outside its proper element, then an explanation was required.

Now, the explanation for why a body was in motion could be that it was going to rest at its proper place; this was the case of natural motion like rocks and rain falling or smoke rising though air. Or the explanation could be that the body was being made to move by an external agent; this was the case of violent motion like a cart pulled by a horse, or a boat sailing over the water, or rain blown by the wind, or weights being lifted from the ground to the top of a building. However, both natural and violent motions required a force; the only difference was that in natural motion the motive force was internal to the body, whereas in violent motion the force was external. For example, falling bodies fell because of their inherent tendency to go down, namely to go to their natural place if they were not already there; the term *gravity* was used to refer to this internal force, and it was measured by the weight of an object. On the other hand, for a sailboat the wind was obviously the external force, and for a cart the horse.

Sometimes "violent motion" was equated with "forced motion," but in such cases it was understood that by "forced motion" one meant motion caused by an *external* as distinct from *internal* force. Since all motion was forced, the term "forced motion" was sometimes regarded as redundant if taken to mean caused motion, and it was found useful only if taken to mean externally-caused motion. In other words, the term *force* was ambiguous and could mean either any cause of motion or an external cause of motion; this may cause some confusion, but the context usually clarifies the meaning.

All motion, then, whether natural or violent, was caused by a motive force, whether internal or external. There was, however, another condition required by all motion—*resistance*. That is, motion was really the overcoming of resistance. This was so, in part, because all space happened to be filled and there was no vacuum, so that whenever a body was moving it could only move through some medium, be it air, water, oil, molasses, sand, or soil. Even the heavenly region, interplanetary and interstellar space, was not devoid of matter; it was filled with (invisible) aether.

Moreover, it was argued that if there were no resistance to overcome, then a motive force (however small) would make a body move instantaneously, namely with infinite speed; and this was an absurdity since it meant that the body would occupy different places at the same time; it followed that there could not be a void or zero resistance. This argument depended on the idea that speed is inversely proportional to resistance, for this idea would provide the justification of why motion without resistance would be instantaneous; that is, not only was resistance required for motion to occur, but motion was correspondingly slower with greater resistance and faster with lesser resistance.

This quantitative relationship between speed and resistance was apparently taken seriously for the extreme case of zero resistance and used as just indicated in the above argument. However, the relationship was not taken equally seriously for the other end of the spectrum, namely for very strong resistance. That is, when the resistance was very strong, rather than saying that a given force would cause some motion, perhaps at very slow speed, it was held that there was a threshold for motion to occur at all; the force had to be sufficient to overcome the resistance in the first place, and if that was the case then the speed was inversely proportional to the resistance. Here, the typical example was that of a single man trying to pull a ship into dry dock by himself; it is clear that he will not be able to move the ship at all, not even 100 times slower than a team of the one hundred men required to accomplish the task.

The relationship between force and speed (when the resistance was constant) was also sometimes expressed quantitatively. The formula was that at constant resistance the speed is directly proportional to the force. Here the paradigm example was the fall of heavy objects through a fluid like water; heavier objects do sink faster than lighter ones, and do so more or less in proportion to the weight; and weight in this case is the (internal) motive force.

In a modern conceptual framework, and using modern terminology, we could combine the two relationships and obtain the following formula: given that the force can overcome the resistance, the body moves at a speed which is directly proportional to the force and inversely proportional to the resistance, that is, *speed* = (*constant*)x(*force/resistance*).

These ideas had great plausibility and were largely in accordance with observation, except for situations like free fall through air and violent projectile motion. For free fall, the Aristotelian theory implied that a lead ball fell much faster than a rock, so that when dropped from the same height the lead would reach the ground much earlier than the rock; moreover, for a given object its speed of fall should not increase with time because it depended only on its fixed weight and the fixed resistance of the air. The problem of projectiles involved the motion of such things as arrows shot from bows, and the question was where was the force making them move after the projectiles had left the ejector. The Aristotelians were aware of these problems and tried to solve them, but their solutions were found to be increasingly unsatisfactory. However, it was the discussion of these problems that provided one line of development in the rejection of the old physics and the construction of the new one.

2.3 ASTRONOMY

Let us recall the phenomenon of *diurnal motion*: all heavenly bodies appear to revolve daily westward around a terrestrial observer. This is most obvious for the case of the sun, whose rising in the east and setting in the west generates the cycle of night and day; the moon can also be easily seen to do the same; and every night each fixed star appears to follow the same westward trajectory as the previous night.

Since the universe was seen as spherical, this diurnal motion was conceived as the daily rotation of a sphere around a line, called the *axis* of diurnal rotation, which went through the north and the south celestial *poles*, N and S in Figure 2.1 (adapted from Kuhn 1957: 31–36, Harris and Levey 1975: 883); this line also intersected the earth's center and two points on its surface, the north and the south terrestrial poles (N′ and S′). From an

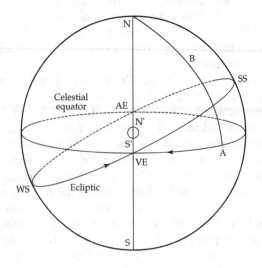

Figure 2.1

observational viewpoint, the celestial poles were the two points in the heavens that appeared motionless (the north celestial pole to observers in the earth's northern hemisphere, and the south celestial pole to observers in the southern hemisphere); and the circular paths of the fixed stars appeared to be centered at the respective poles. On the surface of the celestial sphere, midway between the poles, was a great circle of special importance, called the *celestial equator*; it too had a terrestrial counterpart (the earth's equator), which could be defined as the intersection of the plane of the celestial equator with the earth's surface, or as the great circle on the earth's surface halfway between the north and south terrestrial poles.

One reason for the importance of the celestial poles and equator was that they yielded a fixed frame of reference for the position of heavenly bodies and, correspondingly, for the position of points on the earth's surface. One could measure the angular position of a star north or south of the celestial equator, which was called *declination* (AB, in Figure 2.1); correspondingly, the angular distance from the terrestrial equator of a point on the earth's surface is called *latitude*. For each declination or latitude one could conceive a plane parallel to the equator whose intersection with the surface of each sphere generated circles (called *parallels*) that became smaller as one moved toward a pole. On the other hand, the east-west position of a star (B) required first the drawing of a *meridian*, namely, a great circle (partially shown as NBA) through the star and the poles; then one would measure the *ascension*, namely, the angular distance from this meridian to some particular meridian (for example, A to VE); it was analogous for positions on the earth's surface, except that this east-west angular distance is called *longitude*.

There were two kinds of heavenly bodies, called *fixed stars* and *wandering stars*. A fixed star was a heavenly body that moved daily around the earth in such a way that its position relative to most other heavenly bodies did not change; for example, its declination remained constant, and so did its angular distance from any other fixed star. A wandering star was a heavenly body that not only moved daily around the earth, but also changed its position relative to other heavenly bodies; that is, the wandering stars were those heavenly bodies which, besides undergoing the diurnal motion, appeared to move in other ways (to be discussed presently). There were only seven wandering stars, which were also called planets; indeed the word *planet* originally meant literally "wandering star." Because wandering stars were often called simply planets, fixed stars were often called simply stars. That is, although one broad meaning of the word *star* was synonymous with the term *heavenly body*, one narrow meaning of *star* was identical to the term *fixed star*; the point is that the term *fixed* was often dropped when the context made it clear that one was indeed referring to fixed stars.

Thousands of fixed stars were visible on a clear night with the naked eye; they were catalogued both in terms of apparent brightness (called *magnitude*) and in terms of patterns formed by groups of stars close to each other (called *constellations*). The naked eye could be trained to distinguish six magnitudes; stars of the first magnitude were the brightest, and those of the sixth magnitude were the faintest. The brightest star was named Sirius or the Dog Star; it was located near the equator and was part of the constellation Canis Major. One star of the second magnitude was especially important because it was so close to the north celestial pole that, for practical purposes (such as navigation), it could be regarded to be the pole; it was called Polaris or the North Star and was part of the constellation Little Dipper.

Both the sun and moon were planets because they moved in relation to the fixed stars. Because of their brilliance and their relatively large size, they were called the two *luminaries*. The other known planets were named Mercury, Venus, Mars, Jupiter, and Saturn. We now know that there are other planets circling the sun in orbits beyond Saturn, but they were unknown not only to the ancients but also to Copernicus and Galileo, and so they played no role in the Copernican revolution.

The most basic point about the planets was that, out of the thousands of heavenly bodies, there were seven that circled the earth westward once a day like all others, but did not do so in unison with them; these seven bodies also revolved slowly eastward, so that from day to day their position shifted. That is, whereas a fixed star revolved around the earth in such a way that after twenty-four hours it returned to the same position (relative to other stars) which it had before, after twenty-four hours a planet did not quite return to the earlier position but usually had fallen behind somewhat, being located slightly eastward. This can be seen most easily in the case of the moon by observing its position on succeeding nights at midnight; relative to the fixed stars, it appears to move eastward. The planets seemed to behave as if their motion were a combination of two circular motions in opposite directions: they circled the motionless earth westward with the universal diurnal motion, and in addition they simultaneously moved slowly eastward.

The planets moved eastward at different rates. For example, the moon took about a month to return to the same position relative to the fixed stars; the sun took one year; Mars about two years; and Saturn about twenty-nine years. Thus, the planets moved not only relative to the earth and the fixed stars, but also relative to each other; each planet had its own distinctive motion, besides the universal diurnal motion. Since the westward diurnal motion was common to all, when one spoke of planetary motions one usually referred to the distinctive individual motions of the planets. Note that, while all the individual planetary motions were usually eastward, this direction was opposite to that of the diurnal rotation, which was westward.

The planetary motion of the moon, which took about a month, was the most readily observable one since it was connected with the cycle of its phases; a full moon is easily seen and the period from one full moon to the next is an obvious unit of time that can be used as the basis for a calendar. The planetary motion of the sun was also easy to observe, since it is related to the cycle of the seasons of the year; hence, it was called the *annual motion*. Because of its crucial importance, I shall discuss it in some detail.

Everyone can easily observe that, in the course of a year, the rising or setting sun slowly moves in a north-south direction. Sometimes it rises near due east and sets near due west, which means that it is seen on the celestial equator; this happens around March 21, which is the time of the *vernal equinox*; it also occurs around September 23, the time of the *autumnal equinox*. Sometimes it rises and sets about 23.5 degrees north of due east and due west respectively (namely, north of the celestial equator); this happens in the northern hemisphere around June 22, the time of the *summer solstice*. Sometimes it rises and sets about 23.5 degrees south of due east and due west respectively (south of the celestial equator); this occurs in the northern hemisphere around December 22, the time of the *winter solstice*. One can also observe from a given location on the earth's surface the elevation above the horizon of the sun at noon; in the course of a year this elevation changes daily and ranges about 47 degrees, being highest around June 22 and lowest around December 22 (in the northern hemisphere).

This annual northward and southward motion of the sun indicates that its position relative to the fixed stars changes along a north and south direction since, as stated earlier, the fixed stars remain at a constant distance from the celestial equator. In short, the declination of the sun changes by about 47 degrees during a year, while the declination of a fixed star does not change; so this north-south motion of the sun is part of its wandering among the fixed stars.

Though this apparent solar motion was the one most easily observed, it was not exactly identical to its planetary motion mentioned earlier; for the latter was eastward, whereas the former was northward and southward. The two were related as follows. The sun's eastward revolution in its planetary orbit did not take place in the plane of the celestial equator but in a plane inclined to it by 23.5 degrees. The point was that the sun's motion among the fixed stars was not *exactly* eastward, but *mostly* eastward; its trajectory was slanted north and south. The sun moved eastward and southward for six months, and eastward and northward for the other six months. This can be made clear by means of our diagram, but before explaining it, let us mention a simple kind of observation to detect the sun's eastward motion.

The difficulty in observing the sun's eastward motion among the fixed stars stems from the fact that they cannot be seen when it is visible. What one can do is to observe some star located near the celestial equator and rising in the east soon after the sun sets in the west; this means that the sun and star are diametrically opposed, or about 180 degrees apart. Observe the position of the same star just after sunset about a month later; it will be seen to be not just rising, but high in the sky and about 30 degrees west of its previous position; that means that the sun is now only about 150 degrees away, which is to say that sun has moved eastward about 30 degrees closer to the star. About six months after the first observation, the star will appear and immediately set in the west just after sunset. Twelve months later, the star will again rise in the east when the sun sets in the west.

The planetary motion of the sun may be pictured as in our previous diagram of the celestial sphere (Figure 2.1). Imagine looking at it from above the north celestial pole, and picture the large sphere rotating clockwise around the small motionless central sphere, to represent the westward diurnal rotation of the stellar sphere around the earth. Next, imagine a great circle on the stellar sphere in a plane cutting the equatorial one at an angle of 23.5 degrees, to represent the sun's geocentric orbit projected onto the stellar sphere; in accordance with standard terminology, let us use the term *ecliptic* to refer to this geocentric orbit, or the corresponding great circle on the stellar sphere, or the plane on which they both lie. The intersection of the ecliptic and the equator on the stellar sphere defines two special points, called the vernal equinox (VE) and the autumnal equinox (AE); and halfway around the ecliptic between the equinoxes are two other special points, the summer solstice (SS) at the northern end, and the winter solstice (WS) at the southern end; these four points thus divide the ecliptic circle into four equal quadrants. Now, imagine the sun moving counterclockwise around the ecliptic at a rate that makes it traverse the whole circumference in one year; then the sun will be at VE around March 21, at SS around June 21, at AE around September 21, and at WS around December 21.[3]

Let us now combine the clockwise rotation of the whole stellar sphere with the counterclockwise revolution of the sun along the ecliptic. The result is that the sun, in reality, moved in a helical path which in one year looped clockwise around the earth about 365 times (days of the year), but which in any one day corresponded almost, but not quite, to one of the parallels on the stellar sphere. I say "almost, but not quite" first because the parallel circle was not completely traversed by the sun, but fell short by about one degree (1/360 of a circle, which approximately equals 1/365 of a year); and second because the end of the daily path rises northward or drops southward relative to the beginning of the same daily path by 1/4 of a degree on the average (namely 23.5 degrees every 3 months, or 23.5 degrees every 90 days).

This ecliptic was important also because the other planets are never seen to wander much away from it; that is, planets are always observed to be somewhere inside a narrow belt extending 8 degrees above and below the ecliptic. This was the result of the fact that the individual circular paths of the planets took place in planes which, while not identical with the ecliptic, intersected it at small angles no larger than 8 degrees. This narrow belt on the stellar sphere along which the planets revolved was called the *zodiac*. It was subdivided into 12 equal parts of 30 degrees each, and each part happened to be the location of a group of stars that seemed to be arranged into a distinct pattern. These twelve patterns were the constellations of the zodiac and were named: Aquarius, Pisces, Aries, Taurus, Gemini, Cancer, Leo, Virgo, Libra, Scorpio, Sagittarius, and Capricorn. The sun, moon, and other planets were at all times found somewhere in one of these constellations, and they moved from one constellation to the next in the order just listed.

When projected onto the stellar sphere, the eastward motion of the planets could be described in terms of great circles on the surface of that sphere, all of which were within the zodiac and intersected one another at small angles. But the planets were not believed to be attached to the stellar sphere like the fixed stars; unlike the fixed stars, the planets were not regarded to be equidistant from the earth. The fact that the planets appeared to move relative to the fixed stars, and that this motion took place at different rates for different planets, implied that each planet was attached to its own sphere which rotated eastward at its own rate while being carried westward daily by the diurnal rotation of the stellar sphere. Except for the moon (whose distance was relatively accessible because of eclipses), there was no direct way to measure the sizes of the various planetary spheres or orbits (namely, the distances of the various planets from the earth), but the relative determination was done on the basis of the length of time required for a given planet to complete one circular journey among the stars. The principle used was that the bigger ones of these nested, planetary spheres rotated at slower rates, and the smaller ones at faster rates; that is, the bigger the orbit, the slower the period of revolution. This principle was combined with the observation that the periods of revolution ranged from one month for the moon to one year for the sun and twenty-nine years for Saturn. The result was that in order of increasing distance from the earth, the planets were most commonly arranged as follows: moon, Mercury, Venus, sun, Mars, Jupiter, and Saturn.

One other point about the planets is needed to complete, but also to complicate, this basic sketch of the geocentric universe. Careful observation revealed that no planet moved at a uniform rate in its orbit, but that its speed appeared to vary. Second, although the sun and moon always moved eastward in their planetary revolutions, the other five planets periodically were seen to slow down, stop, and reverse their course and briefly move westward relative to the fixed stars; this reversed movement was labeled *retrogression* or *retrograde* planetary motion. Finally, it was also observed that during retrogression, planets appeared brighter, as if they were nearer the earth than at other times. These observations meant that a planet could not simply be attached to a sphere rotating eastward, for in that case neither the distance nor the direction of revolution nor the speed would ever change. The device most commonly used to explain retrograde motion and variation in brightness and speed was a mechanism consisting of so-called *deferents*, *epicycles*, and *eccentrics*.

The details of this explanation are best postponed until later, when we come to the passage in the *Dialogue* where Galileo discusses retrograde planetary motions (IIIC2). The same applies to other points that are also more technical than the fundamentals sketched in this chapter. For now, let me end this sketch by stressing a very important point: the geostatic and geocentric system of Aristotle, as elaborated by the Ptolemaic system of deferents, epicycles, and eccentrics, yielded plausible explanations and useful predictions; in short, it worked. For about two thousand years, no one was able to come up with anything better. All this changed with Copernicus, to whom we turn in the next chapter.

NOTES

1 The term geostaticism should not be construed in an essentialist manner, but rather in a nominalist manner, which gives it the meaning just defined here. The same applies to all the analogous terms used in this book: geocentrism, heliocentrism, geokineticism, Copernicanism, Aristotelianism, realism, instrumentalism, Catholicism, Protestantism, etc. This caveat is meant to clarify my position on the problem of the undesirable connotations carried by such terms, which has led some scholars (e.g., Westman 2011: 20) to altogether avoid the term "Copernicanism" and similar ones.

2 My account has been inspired by Galileo's *Treatise on the Sphere, or Cosmography*, a short elementary textbook of traditional geostatic astronomy which he wrote and used in the early part of his teaching career, but never published; cf. Galilei 1890–1909, vol. 2: 205–55. My account also relies on Cohen 1960, Kuhn 1957, Lindberg 1992, Rosen (1959, 1992), Toulmin and Goodfield 1961.

3 In modern astronomy, these dates are respectively March 21, June 22, September 23, and December 22, to reflect the fact that the earth's orbit is an ellipse with the sun at a focus, so that the equinox and solstice points do *not* divide that orbit into four equal parts.

3

HISTORICAL CONTEXT

3.1 COPERNICUS'S INNOVATION

In 1543, Copernicus published his epoch-making book, *On the Revolutions of the Heavenly Spheres*. In it he elaborated the details of a world view that may be sketched as follows.[1]

The earth was still spherical and the universe was still finite and spherical. The fixed stars were still attached to the stellar, or celestial, sphere and equidistant from the center. However, the stellar sphere was motionless and did not revolve around the earth with westward diurnal rotation; instead, the diurnal rotation belonged to the earth, though its direction was eastward in order to result in the observational appearance of the whole universe rotating westward. To stress this feature of the Copernican world view (namely, that the earth moves), it may be labeled *geokinetic*.

Moreover, the earth was given a second motion, an orbital revolution around the sun with a period of one year, and also in an eastward direction. In other words, the annual motion was shifted from the sun to the earth (with the direction remaining unchanged), thus making the earth rather than the sun a planet. This terrestrial orbital revolution meant that the earth was located off center, the center being instead the sun; to stress this feature, the Copernican world view may be labeled *heliocentric*.

The moon remained a body that circles the earth eastward once a month. The other five planets continued to be planets, but their orbits were centered on the sun rather than the earth. Around the sun there thus revolved six planets in the following order: Mercury, Venus, Earth, Mars, Jupiter, and Saturn.

Copernicus was updating an idea—the earth's motion—that had been advanced in various forms by the Pythagoreans, Aristarchus, and other astronomers in ancient Greece, but had been almost universally rejected. In fact, during the Copernican revolution, this ancient origin of the geokinetic idea was recognized by using the label *Pythagorean* to refer to it.

Copernicus's accomplishment was to give a *new* argument supporting this *old* idea. His theory was not primarily based on new observational evidence, but was essentially a novel and detailed reinterpretation of available data. He showed in *quantitative* detail that the known facts about the motions of the heavenly bodies (especially the planets) could be explained if the sun rather than the earth is assumed to be at the center, and the earth is taken to be the third planet circling the sun yearly and spinning daily on its own axis. Furthermore, this explanation was simpler, more elegant, more systematic (or coherent), and more powerful than the geostatic explanation.

For example, from the viewpoint of simplicity, there are thousands fewer moving parts in the geokinetic than in the geostatic system. The apparent daily motion of all the heavenly bodies around the earth is explained by the earth's axial rotation, and thus there is only one thing moving daily—the earth—rather than thousands of stars. Hence, insofar as simplicity depends on the number of moving bodies, the geokinetic system is simpler than the geostatic.

A similar point can be made with regard to the number of directions of motion; fewer are needed in the Copernican than in the Ptolemaic system. However, this feature is perhaps more of a question of elegance than of simplicity, because in the geostatic system there are *two* opposite directions of motion, whereas in the geokinetic system all bodies rotate or revolve in the same direction. That is, in the geostatic system, while all the heavenly bodies revolved around the earth with the diurnal motion from *east to west*, the seven planets also simultaneously revolved around it from *west to east*, each in a different period of time. On the other hand, in the geokinetic system, there is only one direction of motion, *west to east*. To see this, consider the following.

Observation reveals a mixture of directions of motion: apparent diurnal motion is westward, but apparent annual motion is eastward. In the Ptolemaic system, basically, appearance corresponds to reality, and so the same mixture is postulated to exist in reality. But in the Copernican system, when apparent westward diurnal motion is explained by means of the earth's axial rotation, the direction has to be reversed, and the earth must be taken to rotate eastward; this is needed because we are replacing a motion in an orbit that encloses the earth with a motion of an earth that is not in an orbit about something else, but is rather spinning around itself. On the other hand, Copernicanism does not have to reverse the direction of the annual motion, because the eastward direction of the apparent annual motion of the sun corresponds to the order of the constellations of the zodiac, and given that the sun appears to move among them in this order, this appearance can only be caused by the earth orbiting in the same direction instead.

With regard to explanatory systematicity (or coherence), this concept means that Copernicus was able to explain many phenomena in detail by means of his basic assumption of a moving earth, without having to add artificial and *ad hoc* assumptions; the phenomena in question were primarily the various known facts about the motions, orbits, and periods of the planets. On the other hand, previously, in the geostatic system, the thesis of a motionless central earth had to be combined with a whole series of unrelated assumptions in order to explain what is observed to happen. The best example of explanatory coherence is the Copernican explanation of retrograde planetary motion and of planetary variation in brightness. This is discussed by Galileo in the Third Day of the *Dialogue* (IIIC2), and we shall examine it then.

Finally, with regard to explanatory power, this generally refers to the number and variety of phenomena which a theory can explain. The Copernican theory had superior explanatory power, partly in the sense that it could explain some features of planetary motion that were mere unexplained coincidences in the Ptolemaic system. Details and examples are beyond the scope of this chapter, but suffice it to say that these were well known facts about various relationships among the motions of the heavenly bodies.

3.2 COPERNICAN CONTROVERSY

As we have seen, the Copernican theory had several advantages over the Ptolemaic one, from the points of view of explanatory simplicity, elegance, coherence, and power. But despite these merits, as a proof of the earth's motion, Copernicus's argument was far from conclusive; and even from the point of view of *confirming* the earth's motion, the argument left something to be desired. The reason for this is that his argument is a hypothetical one, i.e., based on the claim that *if* the earth were in motion *then* the observed phenomena would result. But from this it does not follow with logical necessity that the earth is in motion; all we would be entitled to infer is that the earth's motion offers an explanation of observed facts. Given the merits just mentioned, we could add that they provide some reason for preferring the geokinetic idea, but it is not a decisive or strong reason. It would only be so in the absence of reasons for rejecting the idea; but there were plenty of counterarguments.

The arguments against the earth's motion could be classified into various groups, depending on the discipline or principle involved. In fact, these objections reflected the various traditional beliefs that seemed to contradict the Copernican system. Thus, there were philosophical, epistemological, metaphysical, religious, biblical, theological, physical, mechanical, astronomical, cosmological, and empirical objections. All these are examined in the *Dialogue*, except for the religious arguments, which are criticized in other Galilean writings. Thus, here the nonreligious arguments will be merely mentioned and catalogued, whereas the religious ones will be elaborated more extensively.

Let us begin with the viewpoint of physics, mechanics, or the science of motion. It was argued that the earth's axial rotation and orbital revolution were physically impossible because they contradicted various principles, such as: that the natural motion of terrestrial bodies (consisting of the elements earth and water) is straight toward the center of the universe; and that a simple body (like the earth) can have only one natural motion. It was also argued that if the earth rotated, terrestrial bodies could not move as they are observed to move. For example, freely falling bodies would be slanting to the west; westward gunshots would range farther than eastward gunshots; point-blank gunshots would hit low toward the west and high toward the east; and loose bodies on the earth's surface would be scattered toward the heavens by the extruding power of whirling (i.e., by centrifugal force).

From the standpoint of astronomy—the science of heavenly bodies—the objections were also two-fold. At the general level, if the earth revolved around the sun, it would be the third planet, and hence one of the heavenly bodies, and hence it would share with them many other physical properties; and this contradicted the earth–heaven dichotomy. Moreover, if the earth circled the sun, various phenomena would have to be observed which were not, in fact, observed (until the telescope): the apparent diameter of Mars would undergo periodic changes by a factor of eight; the apparent diameter of Venus would change periodically by a factor of six; Venus would also exhibit a cycle of phases similar to those of the moon; and the fixed stars would undergo annual changes in apparent brightness, diameter, and position.

The earth's motion was also a problem from the viewpoint of epistemology. For Copernicus did not claim that he could feel, see, or otherwise perceive the earth's motion by means of the senses. Like everyone else's, Copernicus's eyes and kinesthetic sense told him that the earth is at rest. Thus, if his theory were true, then the human senses would be failing to detect some motion; and so they would not be telling us the truth, but would be lying to us. But it seemed absurd that the senses should deceive us about such a basic phenomenon as the state of rest or motion of the terrestrial globe on which we live. That is, the geokinetic theory seemed to be in flat contradiction with direct sense experience, and to violate the fundamental epistemological principle claiming that, under normal conditions, the senses provide us with an access to reality.

Finally, there were the religious objections.[2] One of these appealed to the authority of the Bible and may be labeled the *biblical* or *scriptural objection*. It claimed that the idea of the earth moving is heretical because it conflicts with many biblical passages which state or imply that the earth stands still. The most commonly cited such passage was the one reporting God's miracle of stopping the sun in answer to Joshua's prayer: "Then spake Joshua to the Lord in the day when the Lord delivered up the Amorites before the children of Israel, and he said in the sight of Israel, 'Sun, stand thou still upon Gibeon; and thou, Moon, in the valley of Ajalon'. And the sun stood still, and the moon staid, until the people had avenged themselves upon their enemies" (Joshua 10:12–13).

The biblical objection had greater appeal to those who took a literal interpretation of Scripture more seriously (Protestants). However, for those less inclined in this direction (Catholics), the same conclusion could be reinforced by appeal to the *consensus of the Church Fathers*; these were the saints, theologians, and churchmen who had played a formative role in the establishment of Christianity. The argument claimed that all the Church Fathers were unanimous in interpreting relevant biblical passages in accordance with the geostatic view; therefore, the geostatic system is binding on all believers, and to claim otherwise is heretical.

A third theological sounding objection was based crucially on the idea that God is all-powerful, and it may be labeled the divine-omnipotence argument.[3] Its most famous proponent was Pope Urban VIII; it was, in fact, his favorite anti-Copernican objection. This argument is stated, without criticism, at the end of the *Dialogue* (IVA). A version of the argument claimed that since God is all-powerful, He could have created any one of a number of worlds, for example one in which the earth is motionless; therefore, regardless of how much evidence there is supporting the earth's motion, we can never assert that this must be so, for that would be to want to limit God's power to do otherwise. This version of the objection is instructive because it makes it clear that the argument is not purely theological, but also raises issues of a logical, methodological, and epistemological nature. Indeed some such version is essentially correct; for example, part of the argument seems to suggest that scientific knowledge of the physical world is contingently true rather than necessarily true. On the other hand, the argument may also be taken to suggest the possibility that God might be deceiving the human senses and mind; or a general skeptical doubt about physical theories; or a specific difficulty for the Copernican theory; and these suggestions are controversial and questionable.

Copernicus was aware of these difficulties, and this awareness was an important reason why he delayed publication of his book until he was almost on his deathbed. The difficulties were also generally known, and that is the main reason why the geokinetic idea attracted so few followers. However, his argument was so important that it could not be ignored, and various attempts were made to come to terms with it.[4]

Some tried to exploit the mathematical advantages of the Copernican theory, without committing themselves to its physical claims, by adopting an instrumentalist interpretation; that is, the earth's motion was regarded as just a convenient instrument for making mathematical calculations and astronomical predictions, and not a description of physical reality. This instrumentalist interpretation was popularized by an anonymous foreword, preceding Copernicus's own preface, in the printed *Revolutions*, but it was written and inserted without his knowledge by one of the editors supervising the book's publication; he was the German Lutheran pastor, Andreas Osiander (1498–1552). On a different note, the Danish astronomer, Tycho Brahe (1546–1601), undertook an unprecedented effort to systematically collect new observational data, and he also devised another compromise, a theory that was partly heliocentric and partly geocentric, although fully geostatic: the planets revolved around the sun, but the sun moved daily and annually around the motionless central earth. On the other hand, the Italian philosopher and theologian, Giordano Bruno (1548–1600), undertook a multi-faceted defense of Copernicanism that addressed epistemological, metaphysical, theological, and empirical issues; but his defense embodied more of a muddled confusion than a proper synthesis of such issues, and in any case it remained largely unknown. Finally, the German astronomer and mathematician, Johannes Kepler, accepted Copernicanism for metaphysical reasons and then undertook a research program to prove its empirical adequacy, by meticulously analyzing Tycho's data; the result was an improvement of the Copernican theory such that the planets (including the earth) revolved around the sun in elliptical, rather than circular, orbits.

Galileo did not find any of these approaches to the Copernican controversy congenial or sound. However, he did devise his own response, which may be sketched as follows.

3.3 GALILEO'S PURSUIT OF A COPERNICAN RESEARCH PROGRAM

Galileo was born in Pisa in 1564. His father, Vincenzio, was a well known and highly controversial musical practitioner and theorist. One of his important contributions to musicology was to systematically study the sounds produced by strings of various materials, lengths, diameters, and tensions, in order to test various musical theories; he thus influenced Galileo's own experimental approach. In 1581, Galileo enrolled at the University of Pisa to study medicine, but soon switched to mathematics, which he also studied privately outside the university. In 1585, he left the university without a degree and began several years of private teaching and independent research. In 1589, he became professor of mathematics at the University of Pisa, and then, from 1592 to 1610, he held the same position at the University of Padua.

In his early career (1589–1609), Galileo's stance toward the Copernican theory was one of *indirect pursuit*, an attitude that is not only weaker than *acceptance* but also weaker than *direct pursuit*.[5] In fact, during this period his research focused on the general physics of motion rather than on astronomy and cosmology. He was critical of Aristotelian physics, and instead modeled himself on the statics and mathematics of Archimedes of Syracuse (287–212 B.C.), whose work exerted a constant and profound influence on him. Galileo also followed the experimental approach he had learned from his father, using it instead to study the speeds acquired and the distances traversed by bodies falling in various directions during various times.

It was during that period that Galileo formulated, justified, and to some extent systematized various concepts and principles of a new physics: the principles of conservation and composition of motion; and the laws of freely-falling bodies, the pendulum, and motion on inclined planes. However, he did not publish any of these results during that earlier period. Indeed, he did not publish a systematic account of them until the *Two New Sciences* (1638). However, as we shall see later, he did include summaries and previews of these laws and principles in the *Dialogue*, at various points in the discussion when they could be connected with the arguments for and against the motion of the earth.

A main reason for this delay was that in 1609 Galileo became actively involved in astronomy. He was already acquainted with Copernicus's theory and appreciated the importance of his argument. Galileo also had intuited that the geokinetic theory was more in accordance with his new physics than was the geostatic theory. In 1609, he heard about an optical instrument, invented in Holland the year before, consisting of an arrangement of lenses that magnified images three to four times. Without a prototype in his possession, he was soon able to duplicate the instrument, mostly by trial and error. He was also able to increase its magnifying power first to 9 times, then to 20 times, and, by the end of the year, to 30 times. Moreover, rather than merely exploiting the instrument for practical applications on earth, he started using it for systematic observations of the heavens, to learn new truths about the universe.

By its means he made several startling discoveries, which he immediately published in *The Sidereal Messenger* (1610): the moon's surface is rough, full of mountains and valleys; innumerable other stars exist besides those visible with the naked eye; the Milky Way and the nebulas are dense collections of large numbers of individual stars; and Jupiter has four moons revolving around it at different distances and with different periods.

For Galileo personally, the result was that he became a celebrity; he resigned his professorship at Padua; he was appointed Philosopher and Chief Mathematician to the grand duke of Tuscany; and he moved to Florence the same year.

Other discoveries quickly followed (cf. Palmieri 2001, Galilei and Scheiner 2010). In 1610–11, he observed that the apparent disk of Venus, in the course of its orbital revolution, changes regularly from full, to half, to crescent, and back to half and full, in a manner analogous to the phases of the moon. This discovery was announced in private correspondence. And in the next few years, he discovered that the surface of the sun is dotted with dark spots that are generated and dissipated in a very irregular fashion and have highly irregular sizes and shapes, like clouds on earth; but that while they last, these spots move regularly in such a way as to imply that the sun rotates on its axis with a period of about one month. These observations were discussed in *History and Demonstrations Concerning Sunspots* (1613).

Although most of these discoveries were also made independently by others, no one understood their significance as well as Galileo. Methodologically, the telescope implied a revolution in astronomy, insofar as it was a man-made instrument for the gathering of a new kind of data, vastly transcending the previous reliance on naked-eye observation. Substantively, these discoveries provided new observational evidence in favor of the geokinetic hypothesis, as well as refutations of the astronomical objections against it.

For example, lunar mountains and sunspots showed that there are significant similarities between the earth and the heavenly bodies. This refuted the traditional doctrine of the earth–heaven dichotomy; and so it became possible for the earth to be a planet, located in "heaven." Jupiter's satellites showed that it was physically possible for one body to revolve around another, while the latter revolved around a third; and hence it became possible for the earth to revolve around the sun while the moon revolves around the earth. The phases of Venus proved the heliocentricity of its orbit, thus confirming this particular element of Copernicanism. And the multitude of new stars visible with the telescope suggested that fixed stars are not all embedded on the celestial sphere located at the same distance from us, but that their distance varies and is much greater than previously imagined.

Thus, the new telescopic evidence led Galileo to a re-assessment of the status of the geokinetic hypothesis. He now believed not only that the geokinetic theory provided a better explanation of previously known facts than geocentrism (as Copernicus had shown); not only that it was mechanically more adequate (as he himself had been discovering in the twenty years of his early research); but also that it was observationally more accurate in astronomy (as the telescope now revealed). Thus, for the next several years (1609–16), Galileo's attitude toward Copernicanism was one of *direct pursuit* and *tentative acceptance*, by contrast with the *indirect pursuit* of the pre-telescopic situation, and with the *settled acceptance* toward which he was moving but never really reached.

That is, whereas before 1609 Galileo judged that the anti-Copernican evidence outweighed the pro-Copernican one, afterwards he judged the reverse to be the case. However, he also realized that, for various reasons, this change in the probative status of the Copernican theory was not equivalent to settling the issue. There was still some astronomical counterevidence: mainly, the lack of annual stellar parallax. Moreover, he had published nothing of his new physics, and so the physical objections to the earth's motion had not yet been explicitly refuted, and the physics of a moving earth had not yet been articulated. Furthermore, the epistemological objections had also not yet been explicitly answered but, on the contrary, they had acquired a new urgency in light of the new problem about the role, legitimacy, and reliability of artificial instruments like the telescope. Thus it was, during this period, that Galileo conceived a work on the system of the world that would discuss all these aspects of the controversy, and thus provide a confirmation of the earth's motion, as well as a synthesis of astronomy, physics, and epistemology.

However, this plan did not come to fruition for another twenty years. This delay was due to the fact that the theological aspect of the controversy got Galileo into trouble with the Inquisition, acquiring a life of its own that drastically changed his life.

3.4 THEOLOGICAL CONTROVERSY AND CONDEMNATION OF COPERNICANISM

As it became known that Galileo was convinced that the new telescopic evidence rendered the Copernican theory of the earth's motion a serious contender for real physical truth, he came increasingly under attack from conservative philosophers and clergymen in Florence.[6] They started arguing that Galileo was a heretic because he believed in the earth's motion and the earth's motion contradicted Scripture.

Although Galileo was aware of the potentially explosive nature of this particular issue, he felt he could not remain silent, but decided to refute the argument. Because of the circumstances of the attacks and to avoid scandalous publicity, he wrote his criticism in the form of long private letters: in December 1613, to his former student Benedetto Castelli, a Benedictine monk and professor of mathematics at Pisa, and, in spring 1615, to the Grand Duchess Christina.

Galileo's letter to Castelli circulated widely, and the conservatives got increasingly upset. The situation was exacerbated in January 1615, when Galileo received the unexpected but welcome support of a Carmelite friar named Paolo Antonio Foscarini, who published a book entitled *Letter on the Opinion, Held by Pythagoreans and by Copernicus, of the Earth's Motion and Sun's Stability and of the New Pythagorean World System*. Although this was written in the form of a letter to the head of the Carmelite order, the book was a public document. Moreover, although Foscarini's arguments overlapped with Galileo's, they had a distinct flavor and original emphasis: that the earth's motion was probably true, and that it was compatible with Scripture.

Thus, in February 1615, a Dominican friar named Niccolò Lorini from Florence, filed a written complaint against Galileo with the Inquisition in Rome, enclosing his "Letter to Castelli" as incriminating evidence. Then in March, another Dominican, Tommaso Caccini, made a personal deposition against Galileo with the Roman Inquisition. An investigation was launched that lasted about a year. Other witnesses, named by Lorini and Caccini, were interrogated.

As part of this inquiry, a committee of Inquisition consultants reported that the earth's motion is absurd and false as a matter of natural philosophy, and heretical, or at least erroneous, as a matter of religion and theology. Galileo himself was not summoned or interrogated, partly because the key witnesses exonerated him, partly because his letters had not been published, and partly because his published writings did not explicitly contain either a categorical assertion of Copernicanism or a denial of the scientific authority of Scripture.

However, in December 1615, Galileo went to Rome of his own accord to defend his views. He was able to talk to many influential Church officials and was received in a friendly manner. And he may be given some credit with having prevented the worst, insofar as the Inquisition did not issue a formal condemnation of Copernicanism as a heresy, in accordance with the consultants' report. However, some fateful developments happened.

First, the Inquisition decided to give Galileo a personal private warning to stop defending, and to abandon, his geokinetic views. The warning was conveyed to Galileo by Cardinal Robert Bellarmine, the most influential and highly respected theologian of the time, with whom Galileo was on good terms, despite their philosophical and scientific differences. The exact content, form, and circumstances of this warning are not completely known, but they are extremely complex and controversial. Moreover, as we shall soon see, the propriety of the later Inquisition proceedings in 1633 hinge on the nature of this warning. For now let us simply note that Bellarmine reported back to the Inquisition that he had warned Galileo, and that Galileo had promised to obey.

Second, on 5 March 1616, there was a public decree issued by the Congregation of the Index, the department of the Church in charge of book censorship. The doctrine of the earth's motion was declared false, contrary to Scripture, and a threat to Catholicism; Foscarini's book was condemned and prohibited completely; Copernicus's book was temporarily banned, pending corrections and revisions; and the decree ordered analogous censures for analogous books. However, Galileo was not mentioned at all.

Third, at this time Galileo began receiving letters from friends in Venice and Pisa saying that there were rumors to the effect that he had been tried, condemned, forced to recant, and given appropriate penalties by the Inquisition. In the light of such rumors, of the confusing message in the Index decree, and of the even more confusing circumstances of Bellarmine's private warning, in May 1616, Galileo managed to obtain a signed certificate from Bellarmine clarifying the following: the cardinal declared that Galileo had been neither tried nor otherwise condemned, but rather personally notified about the Index decree, and about the fact that in light of the decree the geokinetic thesis could be neither held nor defended.

For the next several years Galileo did refrain from defending or explicitly discussing the geokinetic theory. He was not motivated to resume the earlier struggle even when, in 1620, an Index decree published the corrections to Copernicus's *Revolutions* promised in 1616. They amounted to rewording or deleting a dozen passages that contained either religious references or else language expressing Copernicus's realist and anti-instrumentalist interpretation of the earth's motion; that is, his view that the geokinetic idea was or could be physically true. So revised, the book would instead convey the impression that the earth's motion was merely a useful hypothesis, i.e., a convenient instrument for making mathematical calculations and observational predictions.

Analogously, the appearance of some comets in 1618 generated widespread and heated discussions about their nature, location, and significance. Galileo was drawn into this controversy and eventually published a book on the topic, *The Assayer* (1623). But the connection with the earth's motion was implicit and indirect at best.

3.5 TRIAL AND CONDEMNATION OF GALILEO

The event that put an end to Galileo's silence took place in 1623, when Cardinal Maffeo Barberini was elected Pope Urban VIII. Urban was a well-educated Florentine and, in 1616, he had been instrumental in preventing the direct condemnation of Galileo and the formal condemnation of Copernicanism as a heresy. He was also a great admirer of Galileo and, in 1620, he had even written a poem in praise of Galileo. Further, at about this time, Galileo's *Assayer* was being published in Rome by the Lincean Academy, and so it was decided to dedicate the book to the new pope. Urban appreciated the gesture and liked the book very much. Thus, as soon as circumstances allowed, in the spring of 1624, Galileo went to Rome to pay his respects to the pontiff; he stayed about six weeks and was warmly received by Church officials in general and the pope in particular, who granted him weekly audiences.

The details of these conversations are not known. There is evidence, however, that Urban did not think Copernicanism to be a heresy, or to have been declared a heresy by the Church in 1616. He interpreted the Index decree to mean that the earth's motion was a dangerous doctrine whose discussion required special care. He thought the theory could never be proved to be true, and here it should be recalled that his favorite argument for this skepticism was the divine-omnipotence objection. Thus, Urban's position seemed to be that, as long as one exercised the proper care, there was nothing wrong with the hypothetical discussion of Copernicanism, i.e., with treating the earth's motion as a hypothesis and studying its consequences, its value for explaining physical reality, and its utility for making astronomical calculations and predictions.

Galileo must have gotten some such impression during his conversations with Urban, for upon his return to Florence he began working on a book. This was the work on the system of the world which he had conceived at the time of his first telescopic discoveries, but it now acquired a new form and new dimensions in view of all that he had learned and experienced since. After a number of delays in its writing, licensing, and printing, the work was finally published in Florence in February 1632, with the title *Dialogue on the Two Chief World Systems, Ptolemaic and Copernican*. The author had done a number of things to avoid trouble, to ensure compliance with the many restrictions under which he was operating, and to satisfy the various censors who issued him permissions to print.

To emphasize the hypothetical character of the discussion, Galileo had originally entitled the book *Dialogue on the Tides* and structured it accordingly. That is, it was to begin with a statement of the problem of the cause of tides, and then it would introduce the earth's motion as a hypothetical cause of the phenomenon; this would lead to the problem of the earth's motion, and to a discussion of the arguments for and against, as a way of assessing the merits of this hypothetical explanation of the tides.[7] However, the book censors, interpreting and acting upon the pope's wishes, decided to make the book look like a vindication of the Index decree of 1616. The book's preface, whose content must be regarded as originating primarily from the pope and the censors and only secondarily from Galileo, claimed that the work was being published to prove to non-Catholics that Catholics knew all the scientific arguments, and so their decision to believe in the geostatic theory was motivated by religious reasons and not by scientific ignorance. It went on to add that the scientific arguments seemed to favor the geokinetic theory, but that they were inconclusive, and thus the earth's motion remained a hypothesis.

Galileo also complied with the explicit request to end the book with a statement of the pope's favorite argument—the objection from divine omnipotence. Moreover, to make sure he would not be seen as holding or defending the geokinetic thesis, the author did two things. First, he wrote the book in the form of a dialogue among three speakers: Simplicio, an expert defending the geostatic side; Salviati, an expert taking the Copernican view; and Sagredo, an uncommitted layman who listens to both sides and accepts the arguments that seem to survive critical scrutiny. And, second, in many places throughout the book, usually at the end of a particular topic, the Copernican Salviati utters the qualification that the purpose of the discussion is information and enlightenment, and not to decide the issue, which is a task to be reserved for the proper authorities. Finally, it should be mentioned that Galileo obtained written permissions to print the book, first from the proper Church officials in Rome (when the plan was to publish the book there), and then from the proper officials in Florence (when a number of external circumstances dictated that the book be printed in the Tuscan capital).

The book was well-received in scientific circles. However, a number of complaints began circulating in Rome. One was that the book did not treat the earth's motion as a hypothesis, because it did not regard it merely as a convenient instrument of calculation and prediction but also as a real possibility; that is, the proposition that the earth moves was regarded as a description of physical reality that could be true or false, even if one could not yet be sure as to which was the case. Another charge was that the book defended the earth's motion because the arguments against it were criticized, but the arguments for it were favorably presented. Both of these points involved an alleged violation of the Index decree and of Bellarmine's warning. But there was a third charge: that the book violated a special formal injunction issued to Galileo in 1616, which prohibited him from discussing the earth's motion in any way whatever; a document recording this special injunction had been found in the file of the earlier Inquisition proceedings of 1615–16.

There were also many other smaller complaints. The sheer number in the whole list and the seriousness of some charges were such that the pope might have been forced to take some action even under normal circumstances. But Urban was himself in political trouble due to his behavior in the Thirty Years War between Catholics and Protestants (1618–48). At that particular juncture he was in an especially vulnerable position, and thus not only could he not protect Galileo, but he probably chose to use Galileo as a scapegoat to reassert, exhibit, and test his authority and power. Thus Galileo was summoned to Rome to stand trial.

After various delays, Galileo finally arrived in Rome in February 1633. He was not jailed in the Inquisition prison (as was the norm), but was allowed to lodge at the Tuscan embassy (Palazzo Firenze), under orders not to socialize and to keep himself in seclusion until he was called for interrogation. The proceedings did not begin until April.

At the first hearing, Galileo was asked about the *Dialogue* and the events of 1616. His deposition contained three main points. First, he admitted receiving from Bellarmine the warning that the earth's motion could not be held or defended, but only discussed hypothetically. Second, he denied receiving a special injunction not to discuss the topic "in any way whatever," and in his defense he introduced Bellarmine's certificate, which only mentioned the prohibition to hold or defend. Third, Galileo also claimed that the book did not really defend the earth's motion, because it discussed the arguments on both sides and suggested that the favorable arguments were inconclusive.

The special injunction must have surprised Galileo as much as Bellarmine's certificate did the inquisitors. In fact, it took three weeks before they decided on the next step. The inquisitors opted for some out-of-court plea-bargaining: they would not press the most serious, but most questionable, charge (violation of the special injunction), but Galileo would have to plead guilty to a lesser and more provable charge (transgression of the warning not to defend Copernicanism). He requested a few days to devise a dignified way of pleading guilty to this lesser charge.

Thus, at later hearings, he stated that the first deposition had prompted him to re-read his book. He was surprised to find that it gave readers the impression that the author was defending the earth's motion, even though this had not been his intention. He attributed his error to wanting to appear clever by making the weaker side look stronger. He was sorry and ready to make amends.

Although the authorities accepted this confession of guilt, they were unsure about Galileo's denial of a malicious intention. Thus, in accordance with standard practice, they decided to subject him to an interrogation under the verbal threat of torture. This occurred on June 21, and the transcript indicates that Galileo was threatened with torture but was not actually tortured, and that he was ready to be tortured, and even die, rather than admit his transgression to have been intentional. Such words and demeanor vindicated the purity of his intention.

The trial ended on 22 June 1633, with a harsher sentence than Galileo had been led to believe he would receive. The verdict found him guilty of a category of heresy intermediate between the most and the least serious, called "vehement suspicion of heresy." The objectionable beliefs were the astronomical thesis that the earth moves, as well as the methodological and theological principle that the Bible is not a scientific authority. He was forced to recite a humiliating "abjuration." And the *Dialogue* was banned.

The sentence also states that he was to be held in prison indefinitely. However, this particular penalty was immediately commuted to house arrest. Accordingly, for about one week he was confined to Villa Medici, a sumptuous palace in Rome belonging to the Tuscan grand duke. Then for about five months he was sent to the residence of the Archbishop of Siena, who was a good friend of Galileo's. Finally, in December 1633 he was allowed to return to his own villa in Arcetri, near Florence, to live there in a condition of house arrest.

One of the ironic results of this condemnation was that, to keep his sanity, Galileo went back to his earlier research on motion, organized his notes, and five years later published his most important contribution to physics, the *Two New Sciences* (1638). Without the tragedy of the trial, he might have never done it.

He died in Arcetri in 1642, assisted and surrounded by family and disciples.

NOTES

1 Here my account relies in part on Kuhn 1957: 160–65, Lakatos and Zahar 1975: 368–81, Millman 1976, Wallis 1952: 528–29. Cf. also Finocchiaro (1989: 15–17, 1997: 28–31, 2010b: 21–24).

2 For more on the biblical objection and the one from the consensus of Church Fathers, see the sources in Blackwell 1991: 217–63, Campanella 1994, Finocchiaro 1989: 49–118 and Galilei 2008: 103–67. Cf. the analysis in Finocchiaro 2010b: 65–96, 243–48.

3 Cf. Beltrán Marí 2006: 412–37, Besomi-Helbing 899–902, Camerota 2004: 406–17, Finocchiaro (1980: 8–12, 1985, 1997: 306–8), Speller 2008: 375–96.

4 Cf. Barker and Goldstein 1998, Bucciantini 2003, Gatti (1999, 2002), Westman 2011.

5 For a defense of this view, and a criticism of alternatives, see Finocchiaro 2010b: 37–64.

6 For more details regarding this section and the next, see Finocchiaro 1989 and Galilei 2008 for the documents; and see Fantoli (2003, 2012) and Finocchiaro (1989: 1–43, 1997: 28–47, 2010b: 137–54) for narrative accounts, with references to the relevant scholarship.

7 For an elaboration of this view and a criticism of alternatives, see Drake 1986, Finocchiaro (1980: 16–18, 76–78; 2010b: 37–64, 229–50) and MacLachlan 1990.

Part II

MAIN ARGUMENT IN THE *DIALOGUE*

4

DAY I
SIMILARITY OF EARTH AND HEAVEN

Keeping in mind the preliminaries of the last three chapters, we are now ready to follow the discussion in Galileo's *Dialogue*. Thus, in the next four chapters, we shall summarize the book's main argument, retaining its explicit division into four "Days." But, as explained in the Preface, within each Day, we shall make more explicit than Galileo does its implicit subdivisions into various broad parts (here designated by capital letters) and into specific sections (here designated by Arabic numerals). The parenthetical numbers in the section titles refer to the pages in the Modern Library edition of Stillman Drake's English translation (Galilei 2001, abbreviated DML). And the Appendix contains cross-references to Drake's University of California edition (Galilei 1967, abbreviated DCA), to the critical edition in volume 7 of Galileo's collected works edited by Antonio Favaro (Galilei 1897, abbreviated FAV), and to my abridged translation (Galilei 1997, abbreviated FIN).

We shall focus on the most important thread of the discussion, i.e., the probative strength of the evidence and arguments for the two chief alternative world systems; and since these are the geostatic and geokinetic world views, that thread can also be described as the earth's motion and the arguments and evidence for and against it. This focus means that we shall mostly disregard substantive digressions, secondary topics, rhetorical points, purely philosophical discussions, and dramatic interactions among the speakers. However, some of these latter features will be analyzed later (in Part III). Moreover, although little or no account will be taken of the distinct identities of the three interlocutors (Salviati, Simplicio, and Sagredo), they will be all regarded as partial representatives of Galileo's own mind; their contributions will be combined from the point of view of the key guiding idea of reasoning about the earth's motion.

Finally, the primary concern will be understanding, interpretation, clarification, and coherent reconstruction of the intellectual content of the text. Although we shall also undertake some evaluation and criticism, this will be a secondary concern here and will be treated more explicitly later (in Part III). Similarly, we shall provide some background and contextualization in the course of this reconstruction, but that will be relatively secondary, for it has been done primarily in the preliminaries provided earlier (Part I).

IA1. PERFECTION AND THREE-DIMENSIONALITY OF THE WORLD (DML 9–15)

The *Dialogue* begins with a discussion of the same topic with which Aristotle starts his book, *On the Heavens* (268a1–b10): the perfection and three-dimensionality of the world. The Aristotelian argument may be reconstructed as follows. The world is perfect because it has all the three dimensions that exist—length, width, and depth. And the world is three-dimensional because the number three has special properties, such as: three is the number of parts that everything has, namely beginning, middle, and end; three is the number of sacrifices we make to the gods; and three is the least number of things required before the word "all" can be applied to refer to them collectively (in contrast to saying "one" or "both").

Galileo finds it difficult to take this argument seriously, but he does offer the following criticism. He agrees that the world is perfect and three-dimensional. However, he does not think that perfection has anything to do with three-dimensionality; rather the reason why the world is perfect is that it is "the chief work of God" (DML 15). Nor does Galileo think that three-dimensionality can be grounded on the special properties of the number three; rather it must be grounded on the fact that only three mutually perpendicular straight lines that can be drawn through any given point. And there is no special perfection about the number three: "I do not understand how or believe that with regard to number of legs three is more perfect than four or two" (FAV 35; cf. DML 11).

IA2. ARISTOTLE'S DOCTRINE OF TWO NATURAL MOTIONS (DML 15–20)

The discussion then moves on to Aristotle's doctrine of natural motions (*On the Heavens*, 268b11–269a9), which Galileo criticizes as conceptually incoherent. The doctrine claims that there are two distinct kinds of natural motions: the first is straight, toward or away from the center of the universe, and characteristic of ele-mentary terrestrial bodies such as earth, water, air, and fire; the other is circular, around the center of the universe, and char-acteristic of heavenly bodies made of a fifth substance named aether. This doctrine reflects several ideas. First, there is the idea of simple motion, i.e., motion along a geometrically simple line, which is a line every part of which is congruent with any other part; this yields the two particular cases of straight and circular motion. Second, there is the idea of cosmologically special motions, which yields the two special cases of straight motion toward or away from the center of the universe, and circular motion around that center. Third, there is the idea of motions that are natural for simple bodies, where natural means spontaneous and simple means elementary or non-composite; that is, earth and water naturally move downwards, air and fire upwards, and heavenly bodies around us. That is, the Aristotelian distinction between straight and circular natural motions is part of a conceptual fra-mework that includes the concepts of a geometrically simple line, a finite spherically bounded universe with a center, and the motion naturally undertaken by simple (i.e., non-composite) bodies.

Galileo criticizes this framework because it is arbitrary, ambiguous, and prejudicially oriented toward a pre-determined observational result; these are flaws which I am subsuming under "conceptual incoherence." One objection is that, from the viewpoint of geometrical simplicity, there is a third simple line, namely the cylindrical helix, and yet Aristotle ignores it. Moreover, even if we limit ourselves to straight and circular lines, any straight line is simple regardless of whether it intersects the center of the universe; and similarly, circular motion around any point is simple, even if that point is not the center of the universe. Third, it is arbitrary to equate the spontaneous downward motion of heavy bodies with motion toward the center of the universe, rather than toward the center of the earth; equally arbitrary is to equate the spontaneous upward motion of light bodies with motion away from the center of the universe; and the same applies to the apparent motion of heavenly bodies. It is also arbitrary to equate different simple motions with the natural motions of different simple bodies. Finally, the concept of a simple body is unclear; it could mean a body whose composition is pure and contains no mixture of different elements, or it could mean a body whose natural motion is geometrically simple.

In short, Galileo is charging that Aristotle's concept of natural motion is flawed insofar as he seems to conflate it in turn with geometrically simple motion, cosmologically center-related motion, and the apparently spontaneous motion of elementary bodies. One can indeed make distinctions within these three motions, but one cannot equate the three distinctions with each other, nor do the resulting subtypes of motions correspond with each other.

IA3. A MORE COHERENT CONCEPT OF NATURAL MOTION (DML 20–36)

This criticism is followed by a constructive elaboration: a more coherent concept of natural motion than the Aristotelian one is possible and desirable. That is, Galileo thinks that the difficulties faced by Aristotle's doctrine of two natural motions are so serious that "it would not be amiss to see whether by any chance (as I believe to be the case) we may, by taking another road, follow a more direct and certain path, and by a better application of architectural principles, establish the primary foundations" (FAV 42–43; cf. DML 20). The architectural principles to be applied better are propositions which Galileo shares with Aristotle: that the universe is three-dimensional, perfect, and well-ordered.

On this basis, Galileo proceeds to argue that the only type of natural motion is circular (cf. Copernicus 1992: 16–17). The tip of the iceberg of his argument is this: straight motion cannot be natural, because it cannot be the natural state of integral bodies, and it cannot naturally be perpetual; whereas circular motion can be natural, because it can be the natural state of integral bodies, and it can be perpetual.

The reasons why straight motion cannot be the natural state of integral bodies are the following (DML 21). First, if it were, then integral bodies would be constantly changing place, and if they were doing this then the universe would not be in perfect order. Second, if straight motion were the natural state of integral bodies, they would have a tendency to move through an infinite distance, since a straight line is in principle infinite; but obviously it is impossible to move through an infinite distance.

The reasoning why straight motion cannot naturally be perpetual is as follows. When straight motion is violent, it obviously cannot be perpetual; but when straight motion is nonviolent (i.e., spontaneous or natural), it can be shown that it cannot be perpetual either. This is so because nonviolent, straight motion must be accelerated, and accelerated nonviolent, straight motion cannot be perpetual. To see the latter (DML 36), consider that when a body moves with accelerated, nonviolent, straight motion it must be approaching a place toward which it has a natural inclination, for otherwise it would not be accelerating; but when the body has reached that place, the acceleration must stop. And to see the former (DML 22–23), i.e., why nonviolent, straight motion must be accelerated, consider that nonviolent, straight motion is the simplest way for a body to move from one place to another toward which it has a natural inclination; but in moving from one place to another toward which it has a natural inclination, a body will acquire speed continuously and gradually (since there would be no reason to acquire one degree of speed rather than another); and if a body acquires speed continuously and gradually then its motion is accelerated.

By contrast, circular motion can be the natural state of integral bodies because a natural state of circular motion preserves order (DML 35).

Finally, the reasoning why circular motion can naturally be perpetual is the following (DML 35–36). Acceleration occurs naturally when a body approaches the point toward which it is inclined to move; retardation occurs naturally when a body moves away from the point toward which it is inclined to move; and circular motion around a point toward which a body is inclined can be interpreted as motion such that the body is simultaneously approaching and moving away from that point. It follows that circular motion around a point toward which a body has a natural inclination would be subject simultaneously to acceleration and retardation. Thus, such circular motion would be uniform; hence it could be perpetual.

Besides arguing that the only type of natural motion is circular, and thus that, contra Aristotle, straight motion is not a type of natural motion, Galileo argues for a more positive thesis about straight motion. That is, straight motion is the simplest means of acquiring the natural state of rest at the proper place, or of circular motion. For straight motion is a good means of restoring order out of disorder (DML 21–22); it is also a possible means of creating order (DML 22–23, 32–35); and it is a natural way of acquiring circular motion (DML 23–32).

Let us reflect on this Galilean concept of natural motion. First, although the argument reaches a conclusion that is incompatible with Aristotle's doctrine of two natural motions, the argument is Aristotelian in spirit: it not only utilizes Aristotle's own ("architectural") principles about the universe being well ordered, but also Galileo exploits various Aristotelian ideas, such as the geometrical distinctions between straight and circular lines, between motion toward or away from or around some center of reference, and between motion that is spontaneously undertaken and motion that can last forever. That is, Galileo reaches an un-Aristotelian conclusion on the basis of essentially Aristotelian reasoning from explicitly Aristotelian premises. This type of reasoning may be called *ad hominem* argumentation, following general usage at the time of Galileo, as well as the technical jargon of some present-day philosophers;[1] but this should be distinguished from the currently common notion of *ad hominem* argument as criticizing a claim by attacking the character or circumstances of the arguer rather than the reasons advanced to justify the claim. This feature also implies that this Galilean argument has primarily a critical function, namely criticizing Aristotle's doctrine of two natural motions. Thus, it is unclear to what extent Galileo really accepts this concept of natural motion.

Second, this concept of natural motion claims that the only type of natural motion is circular. This is equivalent to saying that all natural motion is circular. However, this does not mean that all circular motion is natural; nor does it mean that all actual motion is natural. Thus, Galileo is not equating natural motion with either circular motion or actual motion.

Third, obviously the discussion so far has been abstract, conceptual, a priori, and non-empirical. Earlier, the main direct criticism of Aristotle's doctrine of two natural motions was that it is conceptually incoherent and infected with illegitimate empirical considerations. Now, the main merit of the alternative concept of circular natural motion is that it is more conceptually coherent than Aristotle's doctrine; that it is in accordance with the relevant principles, rather than designed merely to produce a pre-determined empirical result. Finally, Galileo's argument for the concept of circular natural motion obviously is free of empirical considerations and follows a principled procedure; this procedure is proper since one is dealing with formulating a conceptual framework.

IA4. GEOCENTRIC ARGUMENT FROM NATURAL MOTION (DML 36–43)

Aristotle's distinction between straight and circular motions was utilized in several interrelated ways. The most obvious one was to provide a foundation for the earth–heaven dichotomy. Another was to justify the geostatic–geocentric thesis. In both cases, since the distinction was a conceptual one, it yielded theoretical or a priori arguments for their respective conclusions. Correspondingly, Galileo's criticism of that distinction was used to undermine these arguments. The focus now is arguments for the geostatic–geocentric thesis.

The a priori argument from natural motion begins with a statement of the relevant portion of Aristotle's doctrine of two natural motions: the natural motion of bodies composed of the elements earth and water is straight toward the center of the universe. The rest goes like this: the whole earth is the collection of all bodies composed of these elements; therefore, the whole earth must be standing still at the center of the universe, because all its parts have moved or will move naturally to be as near that center as possible, and once this has happened the whole collection cannot move naturally to any other point, or in any other way (such as rotating around itself).

This argument has already been undermined since Galileo's criticism in the previous sections has shown that there is no good theoretical reason to accept Aristotle's principle of natural motion for heavy bodies, and so the argument cannot even get started. However, once we grant this principle, the rest of the argument is cogent, and the geostatic–geocentric conclusion is inescapable. Thus, Galileo has criticized this argument at the only point where it is vulnerable.

However, in light of that criticism, one could attempt an empirical re-interpretation of the Aristotelian principle of natural motion. This is precisely what Simplicio does with the following argument:

> who is so blind as not to see that the parts of the element earth and of the element water, being heavy bodies, move naturally downwards, namely toward the center of the universe, assigned by nature herself as the goal and end of straight downward motion; and similarly that fire and air move straight toward the concave side of the lunar orb, as the natural end of upward motion? Since this can be seen clearly, and since we are sure that the same holds for the whole as for the parts, must we not conclude that it is true and manifest that the natural motion of the earth is straight toward the center, and of fire straight away from the center?
>
> (FIN 83–84; cf. DML 36–37, Aristotle 296b7–297a1)

Disregarding the secondary strand about fire and air, and limiting himself to the main strand regarding earth and water, Galileo raises three objections. First (DML 37, FAV 57.20–27),[2] even if there were nothing else wrong with this argument, all it would prove is that if the whole earth were forcibly removed from where it naturally is, it would spontaneously move or try to move back to its natural place with straight motion. But such straight motion would be natural in the sense of spontaneous, not in the sense of potentially everlasting. Hence, such temporary straight motion would not conflict with the circular, natural motion which Copernicanism attributes to the earth, and which would be attributed to it by the doctrine of circular, natural motion elaborated earlier.

Second (DML 37, FAV 57.27–34, 58.5–8), we do not really see heavy bodies spontaneously move straight toward the center of the universe, but rather toward the center of the earth; this is the observational meaning of downwards. It requires further argument to show that this observation implies that they move straight toward the center of the universe. That is, if the key premise of the geocentric argument from natural motion is interpreted as an empirical claim, then although it cannot be refuted by the a priori criticism elaborated earlier, we can legitimately point out that it does not correspond to direct observation and can ask for supporting evidence.

Third (DML 37, FAV 57.34–58.4), there is a difficulty with the argument's second key premise: that the same holds for the whole as for the parts. If taken seriously, it leads to an absurdity. For we have seen that what holds for the parts of the earth is that they spontaneously move straight toward the center of the earth. Then this parts–whole principle implies that it is also true of the whole earth that it spontaneously moves straight toward its own center. But this is an inherent logical impossibility; the whole earth cannot move toward its own center, either spontaneously or forcibly, or by a straight or any other line.

Responding to the second criticism, Simplicio paraphrases the following argument from Aristotle:

> you cast doubt on whether the parts of the earth move in order to go where Aristotle claimed (toward the center of the universe), as if he

had not conclusively demonstrated it by means of contrary motions, when he argues as follows: the motion of heavy bodies is contrary to that of light ones; but the motion of light bodies is seen to be directly upwards, namely, toward the circumference of the universe; therefore, the motion of heavy bodies is directly toward the center of the universe, and it happens accidentally that it is toward the center of the earth because the latter coincides with the former.

(FIN 86; cf. DML 38–39)

This argument attempts to provide empirical evidence for the natural motion of heavy bodies being toward the center of the universe. It is also a cryptic version of the rest of the geostatic–geocentric argument, insofar as it states that the two centers must coincide, as a consequence of the earth parts moving straight toward the center of the universe.

Galileo's criticism is that this argument begs the question: it assumes what it is trying to prove, namely that the centers of the earth and universe coincide. There are two ways to see this. First (DML 40), there is no doubt that light bodies can be seen to move naturally straight upwards. This means that they move directly away from the circumference of the earth and directly toward some circumference greater than, and concentric with, that of the earth, such as that of the lunar orb. But whether the circumference of the universe is one of these greater concentric circumferences toward which light bodies move directly depends on whether the centers of the earth and of the universe coincide. Thus we cannot know that light bodies move directly toward the circumference of the universe unless we first know that the center of the earth is located there. Since one of the argument's premises claims that light bodies move directly toward the circumference of the universe, the argument is assuming what it is trying to prove.

Second (DML 37–38, 40–41), there is another way of showing this. In a spherical, finite universe (being presupposed here), the motion of light bodies is in a sense toward the circumference of the universe, for the simple reason that any straight line anywhere in the universe, if extended indefinitely, will intersect the circumference of the universe. However, the contrary of such a direction is not necessarily toward the center, unless it was to begin with not only (indirectly) toward the circumference but also (directly) away from the center of the universe. Now, the great majority of lines that one can define will lack this property; the only ones that have it are those that intersect the circumference at right angles and go through the center. But we cannot know whether the motion of light bodies is along lines having this property, unless we know that the earth is located at the center of the universe. Thus, again, the argument is assuming what it is trying to prove, and begs the question.

In sum, the geocentric argument from natural motion tries to show that the terrestrial globe stands still at the center of the universe from premises about the natural motion of terrestrial bodies; the argument is groundless if natural motion is formulated as an a priori principle, and it is question-begging if natural motion is formulated as an empirical generalization.

IA5. CONTRARIETY AND CHANGE (DML 43–53)

For the Aristotelians, the distinction between straight and circular natural motions was the key element of the earth–heaven dichotomy, insofar as the distinction per se divided the universe into two regions where bodies moved in accordance with radically different laws. Moreover, the distinction helped to generate other differences between the two regions. The most crucial of these other differences involved the alleged non-occurrence in the heavenly region of any of the physical changes that are constant on earth, such as: generation, destruction, birth, growth, decay, death, weather, and alterations of size, color, and shape. To reach such a conclusion about the unchangeability of the heavenly bodies, the Aristotelians combined the doctrine of two natural motions with another doctrine—about change and contrariety; this held that all change derives, and can only derive, from the action of contraries such as heat and cold, wet and dry, density and rarity, up and down, and light and heavy. Note that, strictly speaking, the Aristotelians regarded motion as a physical change, specifically change of place, and hence as the one and only change that also occurred in the heavens; but in the present discussion the physical changes in question are those other than changes of place. The key argument was succinctly stated by Galileo as follows:

> Generation and corruption occur only where there are contraries; contraries exist only among simple natural bodies, movable in contrary motions; contrary motions include only those made in straight lines between opposite ends; of these there are but two, namely, from the middle and toward the middle; and such motions belong to no natural bodies except earth, fire, and the two other elements; therefore generation and corruption exist only among the elements. And because the third simple motion, namely, the circular, about the middle, has no contrary (because the other two are contraries, and one thing has but one contrary), therefore that natural body to which such motion belongs lacks a contrary and, having no contrary, is ingenerable and incorruptible, etc., because where there are no contraries there is no generation, corruption, etc. But such motion belongs to celestial bodies alone; therefore only these are ingenerable, incorruptible, etc.
>
> (DML 44)[3]

Galileo criticizes this argument in several ways. First (DML 43–44), if it were otherwise correct, the only proper conclusion to be inferred would be: either the earth as well as the heavenly bodies are unchangeable, or the heavenly bodies as well as the earth are changeable, or there is no connection between change and the distinction between straight and circular natural motions. To see this, let us begin by noting that there are two sides to this argument: one argues that terrestrial bodies, and only terrestrial bodies, are changeable; the other argues that heavenly bodies, and only heavenly bodies, are unchangeable. Next, in light of the previous criticism of Aristotle's doctrine of two natural motions and of Galileo's elaboration of a more coherent concept, the natural motions mentioned in the argument from contrariety ought to be conceived either as nonviolent, spontaneous motions or as potentially perpetual motions. Now, if natural motion is conceived as nonviolent, spontaneous motion, then straight motions toward and away from the center of an integral body belong, respectively, to the denser and rarer parts of any such integral body, and what would follow is that the heavenly bodies

as well as the earth are changeable. But if natural motion is conceived as potentially perpetual motion, then (as previously shown)
circular motion would be the only type of natural motion, and so
there would be no contrary natural motions; it would follow that
the earth as well as the heavenly bodies are unchangeable. Finally,
if natural motion is indiscriminately conceived in both ways,
then, from the doctrine of change stemming from contrariety, one
could argue that the heavenly bodies as well as the earth are both
changeable and unchangeable; and this contradiction could be
avoided only by rejecting that doctrine and saying that change
has nothing to do with contrariety, at least with the contrariety of
natural motions.

Second (DML 44–46), the main further consequence of the
argument from contrariety is that the earth stands still at the
center of the universe, since that argument is intended to
be expanded as follows: because heavenly bodies are unchangeable,
and the earth is changeable, the earth is not a heavenly body;
hence the earth does not have the annual motion of revolving
around the sun. However, the argument's main premise (the
doctrine of change stemming from contrariety) is more difficult to
ascertain than the earth's motion or rest; for the earth is a very
large and accessible body, whereas the connection between change
and contrariety is impossible to find in most phenomena; for
example, it is not clear what or where the contraries are in the
spontaneous generation of some insects, in the different rates of
change of most living things, and in transformations resulting
from the transposition of parts.

With this second objection, Galileo is attributing to the contrariety argument an epistemological or rhetorical flaw, rather
than a purely logical one. For this flaw belongs to the bigger
argument of which the contrariety argument is a part, and the
flaw is that the bigger argument tries to prove a conclusion on
the basis of a premise whose truth is harder to know than that
conclusion. And such a procedure is epistemologically self-
defeating in the search for knowledge, and rhetorically ineffective
for the persuasion of someone with a different opinion.

Third (DML 46–48), Galileo objects that the argument from
contrariety is self-contradictory. The reason is that, if one accepts

it, then one could give an analogous argument showing the opposite, namely that the heavenly bodies are changeable. Such an analogous counterargument could be stated as follows: heavenly bodies have contraries, since they are unchangeable, but terrestrial bodies are changeable, and changeability and unchangeability are contraries; now, all bodies that have contraries are changeable, since change does not occur unless there is contrariety, and this means either that (a) change does not occur to a body unless it has a contrary, or (b) change does not occur in a region unless there is contrariety within that region; but the latter alternative (b) cannot be because it would imply that there is no change within the region of the elements earth and water and no change within the region of the elements air and fire, and hence that terrestrial bodies are unchangeable; it follows that heavenly bodies are changeable.

Simplicio counters this criticism by comparing it to the liar's paradox, in particular to the following version. A man from Crete said "I am lying." In so saying, either he was lying, or he was telling the truth. If he was lying, then "I am lying" was a lie; but if "I am lying" was a lie, he was telling the truth; therefore, if the man was lying, he was telling the truth. However, if he was telling the truth, then "I am lying" was true; but if "I am lying" was true, he was lying; therefore, if the man was telling the truth, he was lying.

In this paradox, from each of the two alternative assumptions, one deduces its opposite. However, in the Galilean criticism of the contrariety argument, although one can deduce heavenly changeability from heavenly unchangeability, one cannot deduce unchangeability from changeability. Thus, whereas the liar's paradox generates an infinite series of deductions, the present Galilean counterargument terminates once heavenly changeability has been derived. Moreover, even if the Galilean criticism were analogous to the liar's paradox, that would not undermine the criticism, but rather it would undermine the contrariety argument, because the latter would continue to be the source of the paradox.

Fourth (DML 48–51), the contrariety argument is questionable insofar as it claims that the heavens lack contrariety because they are the domain of natural circular motion, which lacks a contrary. The difficulty is that even if the heavens lack this source of contrariety, they do not necessarily lack other sources. In fact, Aristotle regarded heavenly bodies as the denser parts of the heavens, insofar as the whole heavenly region consisted of the fifth element aether, most of which occurred in ·the relatively rarified form of the invisible heavenly orbs, whereas some of it occurred in the high concentration that generated the bodies of the sun, moon, planets, and fixed stars. But if differences of rarity and density exist in the heavens, then a change-producing contrariety exists there. For differences of rarity and density produces the light–heavy contrariety of terrestrial bodies; this contrariety produces the upward and downward spontaneous motions; and these motions are the source of terrestrial changes (according to Aristotle); hence, differences of rarity and density are the cause of terrestrial changes. Moreover, the cause of terrestrial as well as heavenly differences of rarity and density is the quantitative difference of more or less elementary substance in a given space; and the cause of terrestrial differences of rarity and density is not the qualitative difference of heat and cold, since the density of solid substances changes little when their heat content changes significantly; hence, the cause of terrestrial differences of rarity and density is the same as the cause of heavenly differences of rarity and density. It follows further that heavenly bodies are changeable.

This Galilean criticism is interesting because it is an *ad hominem* argument in the seventeenth-century sense that it derives a conclusion previously not accepted by opponents from propositions accepted by them but not necessarily by the arguer, thus showing that the derived conclusion should be accepted by one's opponents. The conclusion here is that there is some contrariety in the heavens, and the main Aristotelian premise is that in the heavens there are different concentrations of the fifth element, aether. We saw earlier (IA3), that Galileo used this type of argument in proposing a more coherent concept of natural motion than the Aristotelian doctrine of two natural motions.

Finally (DML 51–52), Galileo charges that the argument from contrariety commits an equivocation involving the term "body". That is, when the argument deals with bodies undergoing change due to contrariety, the term "bodies" refers to parts of an integral (whole) body, namely parts of the terrestrial globe; this remains true even if the parts are the elements earth, water, air, and fire, but it is more striking if the parts are individual bodies (like rocks) composed of one of these elements, or a mixture thereof. However, when the argument deals with bodies undergoing no change because of a lack of contrariety, the meaning of the terms "bodies" is that of integral (whole) bodies, namely heavenly globes such as the moon, sun, planet Venus, and North Star. This is illegitimate, for the argument is in effect reaching a conclusion about whole bodies from premises about parts of whole bodies.

If the meaning of the term is kept constant, as it should be, then we could do either one of two things. We could start with some relatively plausible claims about parts of the terrestrial globe being subject to both contrariety and change, and then we could only reach a conclusion about parts of the heavenly globes being similarly subject to contrariety and change. Or we could start with claims about whole integral bodies in the heavens not being subject to either contrariety or change, and then we could reach a conclusion about the terrestrial globe as a whole being no more subject to contrariety and change than the heavenly bodies. However, neither of these arguments would enable us to accomplish what the Aristotelian argument is trying to do.

Thus, besides the earlier flaws, the Aristotelian argument from contrariety commits the fallacy of equivocation. That is, it uses a term ambiguously in such a way as to conceal the fact that if the term is used with one meaning, then the conclusion obviously does not follow; whereas if the term is given the other meaning, then a crucial premise is obviously groundless or false (even if its truth happens to imply the conclusion).

To sum up, contrariety and change were connected by Aristotle by claiming that physical changes derive only from the action of contraries. Starting from this premise, he combined it with the doctrine of two natural motions to arrive at the conclusion that the heavenly bodies are unchangeable. This may be called the anti-Copernican argument from contrariety. Galileo criticizes this argument in five ways; most strikingly, the argument is self-contradictory and commits the fallacy of equivocation.

1A6. OBSERVATION OF HEAVENLY CHANGES (DML 53–66)

The arguments discussed so far have been relatively theoretical. However, the Aristotelians had other, more empirical arguments to justify the earth–heaven dichotomy. The main such argument may be called the observational argument for heavenly unchangeability. Galileo has Simplicio state it as follows:

> sensible experience shows us how on earth there are constantly gen-erations, decay, changes, etc.; none of these have ever been seen in the heavens either with our senses, or according to tradition or the reports of the ancients; therefore, heaven is unchangeable etc., and the earth changeable etc., and thus different from heaven ... On the earth I constantly see the generation and destruction of plants, trees, and animals, and the production of winds, rain, and storms; in short, this aspect of the earth is in a perpetual metamorphosis. None of these changes are seen in the heavenly bodies; their appearance and arrangement have been very exactly the same within human memory, without the generation of anything new or the destruction of anything old.
>
> (FIN 91–92; cf. DML 54)

For a proper appreciation of this argument, it is best to begin with a constructive clarification, advanced by Galileo himself (DML 56–57). That is, the argument as stated is stronger than the one that may correspond more closely to the letter of Aristotle's text. Such a more literal version would read as follows: no one has ever observed any generation or decay of heavenly bodies in the heavenly region; therefore, the heavenly region is unchangeable. This argument would have the flaw of sanctioning the following obviously incorrect inference: no one has ever observed any gen-eration or decay of terrestrial globes in the terrestrial region; therefore, the terrestrial region is unchangeable. What is wrong with the literal version of the Aristotelian argument is that it implicitly contrasts heavenly bodies with such terrestrial bodies as cities, which have been known to come and go, rather than with the terrestrial globe, for which there is no record and no prospect of its appearance or disappearance.

This constructive clarification is a favorable evaluation of the argument under discussion, insofar as it strengthens the argument or defends it from a possible objection. There is another similar constructive clarification, which Galileo does not explicitly make, but which is implicit in his discussion, starting from the way the argument is initially stated.

The argument might be taken to have the form "no As have been observed; therefore, no As exist"; that is, "no heavenly changes have ever been observed; therefore, no heavenly changes exist." This, in turn, might be regarded as tantamount to arguing that because we do not know that something is true, we can conclude that it is false. This is certainly a fallacious manner of reasoning, labeled appeal to ignorance; although it is a common fallacy, it is easily recognized as fallacious when made explicit. Thus, to interpret the Aristotelian argument as an appeal to ignorance would immediately generate this criticism, depriving it of any worth. Such criticism in turn might be criticized as attacking a straw man, namely, as being based on an untenable interpretation of the argument being examined. We might also add that such a criticism would seem to violate the principle of charity, according to which, when evaluating an argument, one should interpret it in such a way that the argument avoids some of the most obvious errors (cf. Scriven 1976: 71–73). Finally, the uncharitable interpretation would also involve taking the passage out of context, for the context makes it clear that the form of the argument is rather the following: "Hs have been observed to have the property not-A (whereas Ts have been observed to have the property A); therefore, Hs have the property not-A (whereas Ts have the property A)"; that is, "heavenly bodies have been observed to be devoid of (qualitative) change; therefore, heavenly bodies are devoid of (qualitative) change."

Moving on to some destructive criticism, a key objection is this (DML 56). The observational argument for heavenly unchangeability is invalid because for an observer on the moon, no terrestrial changes would be noticeable before some particular very large terrestrial change had occurred, and yet terrestrial bodies are obviously changeable and would have been so even before that occurrence.

Here, Galileo is using the "method of counterexample." That is, he envisages a situation similar to the one mentioned in the given argument, involving the observation of the earth from the moon; and he considers the possibility of such an observer arguing that because no terrestrial changes have been observed, therefore the earth is devoid of physical qualitative changes. Now, in the envisaged situation the premise of such an argument would be true, and yet its conclusion would be false. This shows that it is possible for the premises of such an argument to be true while the conclusion is false; hence the conclusion does not follow necessarily from the premises, and the argument is formally invalid. Since the observational argument has the same form as this counterexample argument, it too is formally invalid.

Another way of looking at this criticism is the following. Galileo is saying that the most that would follow from the premise of the observational argument is that there have been no heavenly changes *so far*, i.e., during the period of human observation. However, its conclusion asserts much more than that; hence, as asserted, the conclusion does not follow.

This criticism is correct as far at it goes, but perhaps it does not go very far. For all sides of this discussion clearly realize that this is an a posteriori argument based on sense experience. And they show various degrees of awareness that such empirical arguments cannot establish their conclusions as necessarily true, or beyond any reasonable doubt; such arguments are inductive or merely probable. Perhaps to anticipate this difficulty, Galileo has another criticism, which addresses precisely this issue and interprets the Aristotelian reasoning as an inductive, probable argument (DML 54–56).

The additional criticism is this. If there were changes within the heavenly bodies, most of them could not be observed from the earth, since the distances from the heavenly bodies to the earth are very great, and we know that on earth changes can be observed only when they are relatively close to the observer. Moreover, even if there were changes in the heavenly bodies large enough to be observable from the earth, they might not have been observed, since even large changes cannot be observed unless careful, systematic, and continual observations are made; but no such observations have been made, at least not by the argument's proponents.

This criticism interprets the observational argument for heavenly unchangeability as an inference to the best explanation; that is, it presents the conclusion about heavenly unchangeability as the explanation of the observational absence mentioned in the premise. However, the critic suggests two other ways of explaining the fact mentioned in the premise: it may be due to the great distance between the earth and the heavenly bodies, and/or to the lack of sufficiently careful observations. These alternative explanations do not *refute* the Aristotelian *explanation* but rather the Aristotelian *argument*; that is, this criticism does not prove the conclusion of the original argument false, but rather *weakens* the inferential link between premise and conclusion, for there is no reason to prefer the Aristotelian to the Galilean explanation. Thus, the point being made is a logical criticism, affecting primarily the premise–conclusion relationship in the original argument.

However, Galileo has other objections up his sleeve, advancing more substantive criticism. In fact, his main criticism is that the key Aristotelian premise is false, namely, the claim that no qualitative changes have ever been observed in the heavens. Whatever may have been the case at the time of Aristotle, in Galileo's time changes had been observed both within heavenly bodies and in the heavenly region.

For example, comets were observed in 1618, and they constitute changes in the heavens, regardless of whether they originate there or in the terrestrial region. The novas of 1572 and 1604 are instances of observed generation and decay in the heavens. Sunspots are changes on the body of the sun: for they appear and disappear at random in the solar disk; their apparent motion is slower at the edges and faster in the middle of the solar disk; and their apparent size and shape is narrower at the edges and wider in the middle of the disk; all these observations can be explained only as the effect of the laws of perspective applied to cloud-like phenomena on the surface of a spherical sun rotating around its axis about once a month.

On the other hand, the arguments against the observation of heavenly changes are worthless. For example, the author of the *Anti-Tycho* uncritically assumes that comets are phenomena to which parallax theory is applicable; uncritically rejects observations that do not confirm his view; and is inconsistent in his attitude toward comets and toward novas, since he attempts to locate the former in the terrestrial region, but regards the latter as irrelevant because they do not represent changes within previously known heavenly bodies. And regarding sunspots, various authors claim that they are previously unknown planets circling the sun and obscuring parts of it when they come into our line of sight; but this claim is refuted by the just-mentioned evidence of their random appearance and disappearance and their changes in apparent speed, shape, and size.

Besides refuting the key premise of the Aristotelian argument, Galileo criticizes the argument by refuting its conclusion, and thus justifying its opposite (DML 57–58, 63–65). For in this case, the refutation of the premise is not a purely negative affair: to deny that no changes have ever been observed in the heavens is to affirm that some changes have been observed. But the observation of heavenly changes can be combined with a plausible version of the principle that observation corresponds to reality, to yield the conclusion that the heavenly region is changeable. This principle is not only intrinsically plausible, but also was accepted by the Aristotelians; thus, they too are obliged to accept the thesis of heavenly changeability. In Galileo's own eloquent words:

> I say that in our age we have new phenomena and observations such that, if Aristotle were living nowadays, he would change his mind. This may be obviously gathered from his own manner of philosophizing. For he writes that he regards the heavens as unchangeable, etc. because no one has seen the generation of something new or the destruction of something old; so he implicitly conveys the impression that, if he had seen one of these phenomena, he would have concluded the opposite and given priority to sensible experience over natural theorizing (as is fitting); if he had not wanted to value the senses, he would not have inferred the unchangeability from not sensibly seeing any change.
>
> (FIN 96; cf. DML 57).[4]

IA7. TELEOLOGY AND ANTHROPOCENTRISM (DML 67–71)

The arguments from natural motion, from contrariety, and from observation were not the only reasons the Aristotelians had for holding the earth–heaven dichotomy. Another one was the teleological argument for heavenly unchangeability (cf. Finocchiaro 1980: 35, 109–10, 320–29, 377–79). This was based on the thesis of teleological anthropocentrism: that the whole universe exists for the sake and benefit of mankind. Here, Galileo examines two versions of the argument, first a more confusing one stressing the notion of perfection, and then a more transparent one avoiding this notion. The first version reads as follows:

> these conditions [physical changes] ... would render less perfect the celestial bodies, in which they would be superfluous. For the celestial bodies—that is, the sun, the moon, and the other stars, which are ordained to have no other use than that of service to the earth—need nothing more than motion and light to achieve their end.
>
> (DML 68)

To this argument one could object that it seems to assume that change makes things imperfect, but this assumption is false because change usually makes things *more* perfect. This is so because the less perfect things are usually devoid of change; for example, a desert is less perfect than a garden, and a dead animal is less perfect than a live one. However, when things are both perfect and devoid of change (e.g., precious metals such as gold), they are precious because they are scarce.

However, this objection would not be doing justice to the teleological argument because that argument is not really assuming that change makes things imperfect. In fact, the claim that change makes things imperfect is not explicitly stated in the argument. Nor is the claim implicitly employed as the reason why change would make heavenly bodies imperfect, since the reason is given explicitly, and it is that heavenly changes would be superfluous. In other words, the assumption in question is not being made in the original argument because it contains no implicit or explicit step saying: changes would make heavenly bodies imperfect because changes make things imperfect. Instead the step claims: changes would make heavenly bodies imperfect because heavenly changes would be superfluous.

Moreover, when this objection says that precious metals are precious because they are scarce, this is true only in the sense of "*primarily* because they are scarce," since their relative scarcity is a factor that contributes to their preciousness. But such a proposition is irrelevant. The relevant proposition would be: precious metals are perfect because they are relatively scarce; but this latter proposition is not true because their scarcity has nothing to do with their perfection.

Another objection against the teleological argument might be this. It seems unlikely that the purpose of heavenly bodies is to serve the earth, since this would mean that divine and eternal entities would be serving a base and transitory body. However, this objection would be misconceived. For although various parts of the earth are transitory and subject to change, the earth as a whole is not transitory, but as eternal as the heavenly bodies; thus, if heavenly bodies serve the earth, this means that some eternal bodies are serving another eternal body, and there is no absurdity in that.

Such unfair objections could be more easily avoided if one were to reconstruct the argument quoted above as follows: there cannot be heavenly changes because change would make heavenly bodies imperfect; the reason for this is that such changes would be superfluous; they would be superfluous because heavenly bodies need only light and motion to fulfill their purpose; and this is so because their purpose is to serve mankind.

The real issues emerge more clearly later in this passage, when the argument is formulated so as to avoid any talk of perfection, and to stress more explicitly the teleology and anthropocentrism:

SAGREDO ... why can you not and should you not likewise admit alterations, generations, etc. in the external parts of the celestial globes ... ? SIMPLICIO. This cannot be, because the generations, mutations, etc. which would occur, say, on the moon, would be vain and useless, and *natura nihil frustra facit* [nature does nothing in vain]. SAGREDO. And why should they be vain and useless? SIMPLICIO. Because we plainly see and feel that all generations, changes, etc. that occur on earth are either directly or indirectly designed for the use, comfort, and benefit of man ... Of what use to the human race could generations ever be which might happen on the moon or other planets? Unless you mean that there are men also on the moon who enjoy their fruits; an idea which if not mythical is impious.

(DML 69–70)

This argument may be reconstructed as follows: there cannot be heavenly changes because nature does nothing in vain, and such changes would be vain and useless; for they would not occur for the benefit of mankind, and they would not occur for the benefit of the inhabitants of the heavenly bodies (since there are not such creatures).

Now, although this argument is more powerful, and not subject to the misunderstanding objections discussed earlier, it is vulnerable to three effective critiques.

First, to say that there are no human beings on the heavenly bodies is probably correct. But from this it does not necessarily follow that there are no changes there, since from this same proposition it does not necessarily follow that there are no creatures there. In fact, to think that it does follow would be like arguing, for someone who had lived all his life in a forest, that there could not be things like ships or fish. However, if there are creatures there, then there could be changes for their sake.

Second, the conclusion seems wrong, since some terrestrial changes derive in part from the heavenly bodies, and to cause changes in others without being changed oneself seems unlikely; thus, heavenly bodies, too, probably undergo changes.

Third, to say that the purpose of heavenly bodies is to serve mankind is unjustified, because the only reason one could give is to say that the purpose of everything else is to serve mankind; for example, horses, herbs, cereals, fruits, birds, beasts, and fish exist for the comfort and nourishment of human beings. However, the difficulty with this reason is that at best it shows only that the purpose of all *terrestrial* things is to serve mankind, and from this we cannot conclude that the purpose of heavenly bodies is also to serve mankind (especially for an Aristotelian who distinguishes sharply between the two domains and believes that different principles apply in each).

In sum, the teleological argument for heavenly unchangeability should *not* be interpreted as claiming: physical changes in the heavenly bodies would make them imperfect, because change makes things imperfect; but heavenly bodies are perfect, because the heavenly region is the domain of perfection; therefore, the heavenly bodies do not undergo changes. For, here, the premise that change makes things imperfect cannot be fairly attributed to the Aristotelians, who are aware that in some cases (for terrestrial bodies) change enhances perfection. Instead the argument should be interpreted as follows: physical changes in the heavenly bodies would be superfluous, because they would be of no use to mankind, and the purpose of all things is to benefit mankind; but nature does nothing in vain, and so heavenly bodies do not undergo changes. Here, the premise asserting teleological anthropocentrism is a deep principle of the Aristotelian system, but it can be effectively criticized with strong objections.

IB1. SIMILARITIES BETWEEN EARTH AND MOON (DML 71–82)

So far, the primary purpose of the discussion has been to examine the Aristotelian arguments for the earth–heaven dichotomy. These have been evaluated mostly negatively, although some positive ideas have emerged. The most important of the latter is that the phenomena of comets, novas, and sunspots show that the heavenly region undergoes physical changes similar to those of the terrestrial region. At this point, the primary purpose changes to the presentation of evidence in favor of the similarity between the earth and moon, and hence the similarity between the two regions.

Galileo begins with a summary of several similarities, some of which are later discussed in greater detail. He makes it clear from the start (DML 71) that there are also important differences between the earth and moon, and, in fact, we will find them discussed later.

To summarize, for Galileo, the earth and moon are both spherical in shape, opaque, devoid of their own light, solid, and rough in their surface. Moreover, each body reflects light onto the other; causes eclipses when viewed by observers on the other body; shows periodic changes of shape called phases; and appears unevenly bright.

Then Galileo has Simplicio summarize the Aristotelian disagreements. For some of the Galilean claims, the Aristotelians reject them completely and hold the opposite; for example, Simplicio denies that the lunar surface is rough and holds that it is perfectly smooth. For other claims, Simplicio rejects them only in part and holds a qualified version of the claim; for example, he denies that the moon is completely devoid of its own light, but holds that the faint secondary light shown when the moon is thinly crescent is the moon's own light. For still other cases, Simplicio essentially agrees with the claim's content, but rejects Galileo's supporting reasons and instead gives a different justification; for example, this applies to the claim that the moon is solid.

In at least one case, the two sides agree completely, both with regard to content and supporting evidence. This is the claim that both the moon and earth have a spherical shape (as distinct from being flat or concave). In this case, the supporting evidence consists of observations that can be easily made with the naked eye, even by twenty-first century laypersons. For these reasons, the corresponding passage is of some interest:

> We can be sure that the Moon is similar to the Earth in shape, as there is no doubt that it is spherical: this is proved by the fact we see it as a perfectly circular disc, and by the way in which it receives light from the Sun. If its surface were flat, then it would all be bathed in light at the same moment, and similarly it would all simultaneously become dark; whereas in fact the parts facing the Sun are illuminated first, followed by the rest, so that it is only at full moon, when it is in opposition to the Sun, that the whole disc appears light. Conversely, the opposite would happen if its visible surface were concave; in that case the parts turned away from the Sun would be illuminated first.
>
> (Shea-Davie 185; cf. DML 71–72)[5]

1B2. ROUGHNESS OF THE LUNAR SURFACE (DML 82–100)

Although both sides agree that the moon is a sphere, Galileo hastens to make clear that whereas the Aristotelians hold that the spherical surface of the moon is perfectly smooth, he thinks that the lunar surface is rough and full of mountains and valleys. To show this, Galileo elaborates a multi-faceted argument based partly on four simple experiments of a common-sense sort and partly on careful telescopic observations.

The first experiment is this (DML 82–84). During daylight, while the sun is shining, go into a courtyard and hang a flat mirror on one of the walls, and compare and contrast the ways the mirror and the wall reflect light; you will see that the wall looks brighter than the mirror from all observation points except one spot, where the rays reflected by the mirror all go. The second experiment (DML 86–87) involves hanging a spherical mirror in the courtyard, and comparing and contrasting its reflection with that of the flat mirror; you will see that the spherical mirror has no noticeable effect on the illumination of the objects in the courtyard, and there is no one spot where its reflected rays go, although the observer can see a single ray of light coming from the spherical mirror, regardless of the observation point. These two experiments show that if the moon were perfectly spherical, it would be invisible; for only a single ray of light would reach any one observer on earth, but because of the distance it would be so weak as to be undetectable; thus, the lunar surface must have some roughness.

Next, do a third experiment (DML 92). Fold a sheet of paper into two equal parts, and let one part receive light perpendicularly and the other part obliquely; you will see that the paper lit obliquely appears less bright than the paper lit perpendicularly. Finally (DML 95–96), fold a sheet of paper into two unequal parts, and shine a light perpendicularly on the smaller part and obliquely on the larger; then look at the paper from an angle such that the two unequal parts appear of equal size; you will see that the obliquely-lit paper, although actually larger, still looks less bright than the other smaller paper. These two experiments show that if the lunar surface were not significantly rough but was like a wall or paper, then a full moon would appear brighter at the center and darker near the edges; thus, the lunar surface cannot be smooth like a wall or paper, let alone be perfectly smooth like a mirror.

This argument is based on observations that are noteworthy for their ease and incisiveness, as well as insofar as they involve naked-eye observations. Moreover, they involve experimentation, in the sense of the investigator's intervention into the natural order with the intention of answering some pre-determined question. By contrast, the next argument involves observations that are hard to make and interpret, and rely on an artificial instrument—the telescope. The key observation is that when seen through a telescope, the lunar surface appears mountainous. However, this mountainous appearance is potentially ambiguous, primarily because it could result from the texture of the substance of which the moon is composed; for example, the lunar substance could be similar to mother of pearl, marble, and wood, which can be very smooth and yet appear to the eye to have irregularities.

Here the crucial point is that the alternative explanation has plausibility only at a very general and superficial level. However, we should focus on the specific details of the telescopic appearances, and should utilize an artificial model. The model consists of two opaque balls, a smooth one made of a textured material like marble, and one with a very rough but uniformly colored surface; and they should be subject to illumination from a light source in such a way that the relative positions of the illuminated ball, light source, and observer can change. The use of the model reveals that none of the specific details of the telescopic appearance of the moon can be reproduced with the smooth marble ball, but all of them can be reproduced with the rough colored ball. The specific details are these: during most of the lunar cycle, you can see that part of the lunar disk is lit, and part dark; you can see that the boundary line separating these two parts is not smooth, but very irregular; that there are lighted bulges in the dark region and darker areas in the bright region; that the bulges in the bright region cast shadows; that the length of these shadows changes depending on the distance of the bulge from the boundary line; that the length of the various shadows in the lighted region decreases until there are no more shadows when the whole hemisphere is lit; and that as the darkness starts growing, the same bulges can be recognized and cast shadows whose length increases.

The *Dialogue* contains no drawings illustrating these features, but it is useful to reproduce some here from *The Sidereal Messenger* (Galilei 2008: 52–55):

Figure 4.1

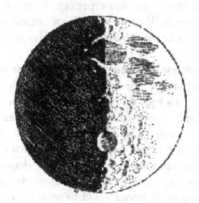

Figure 4.2

A final strand of Galileo's argument consists of a criticism of the Aristotelian reason why the moon, like all other heavenly bodies, is not only spherical, but perfectly spherical. Their reasoning was this (DML 96): the heavenly bodies are perfect in the sense of having all perfections, because they are unchangeable, and unchangeability implies perfection; but the spherical shape is the most perfect; therefore, heavenly bodies are perfectly spherical.

In this passage, Galileo objects to this argument by arguing that if a spherical shape caused unchangeability, all bodies would be unchangeable, because all bodies can be conceived as composed of an indefinite (or infinite) number of smaller spheres; and on Aristotelian principles, such a collection should be unchangeable because there is no contrariety among these constituent spheres, given that they all have the same shape and are made of the same material. Although this argument is plausible per se, ingenious, and *ad hominem* (in the seventeenth-century sense of internal criticism), it is basically irrelevant, because it refutes a claim which is not being made or presupposed by the original argument.

A more relevant objection would have been a criticism of the claim that unchangeability implies perfection, along the lines of Galileo's earlier criticism in the discussion of the teleological argument (IA7). However, in other writings, he criticizes the present argument for perfect sphericity directly, by criticizing the claim about the spherical shape being the most perfect. That criticism deserves quotation because it illustrates Galileo's critical attitude toward the notion of circularity, which in other ways he seems to favor:

> I do not think one can assert absolutely that the spherical shape is more perfect than the others, but rather only relative to some context. For example, for a body that must be able to turn in all directions, the spherical shape is most perfect, and thus the eyes and the head of the femur are made by nature perfectly spherical. By contrast, for a body that must stay motionless and stable, such a shape would be most imperfect, as compared with any other shape; so in building a wall, whoever would use spherical stones would be doing a terrible job, whereas angular stones would be most perfect. If the spherical shape were absolutely more perfect than the others, and if bodies that are more noble had to have more perfect shapes, then the heart and not the eyes should be perfectly spherical, and spherical should have been the liver (which is such a principal organ), rather than some other parts of the body that are most vile.
>
> (Galilei 1890–1909, vol. 11: 146–47; cf. Besomi-Helbing 300)

IB3. MOON'S SECONDARY LIGHT (DML 100–113)

Another similarity between the earth and moon claimed by Galileo regards their optical properties. On the one hand, the moon (like the earth) is dark and not luminous, insofar as it does not give off its own light but only reflects light emanating from other sources (primarily the sun). On the other hand, the earth can reflect light (primarily sunlight) about as well as the moon can. And there is an easily observable phenomenon that ties together these two properties, called the moon's secondary light: the faint light visible with the naked eye in the dark part of the lunar disk when the moon is thinly crescent. This secondary light is to be distinguished from the primary light which derives directly from the sun and generates the crescent and other phases, and which was relatively well understood since Aristotle's time, as due to the changing relative position of the three bodies (sun, moon, and earth). The connection is that the moon's secondary light is the effect of sunlight reflected onto the moon from the earth.

The purpose of this passage is to elaborate an argument supporting this causal claim. Since this claim advances an explanation of the moon's secondary light, and explicitly denies other possible explanations, the supporting argument is an inference to the best explanation. That is, it starts with a well-known fact to be explained—the moon's secondary light. It proposes a possible explanation—light reflected from the earth. It criticizes alternative explanations: light originating from the moon's own body; light coming from other heavenly bodies, chiefly the closest one, Venus; or direct sunlight passing through the translucent lunar body. And it concludes that terrestrial reflection is the best explanation of the moon's secondary light.

The first strand of the argument shows that the earth can reflect sunlight about as well as the moon can (DML 100–103). In fact, sometimes the moon is visible during the day, and appears like a small, whitish patch of cloud against the blue sky, similar to other clouds; this shows the reflecting power of clouds is comparable to that of the moon. Now, sometimes clouds are relatively low and near mountains, and are hard to distinguish from the mountains; this shows that the reflecting power of the solid earth is comparable to that of clouds. Thus, we can conclude that the reflecting power of the solid earth is comparable to that of the moon. Moreover, in some ways moonlight is weaker than light reflected from the solid earth; this can be shown by the phenomenon of twilight at dusk. That is, after the sun has set, the light we see is mostly sunlight reflected from terrestrial bodies such as the ground and mountains; now, even when the moon is visible during twilight, for a while (about half an hour) objects will not cast shadows due to the moonlight, and the cause of this is that the twilight is stronger than the moonlight. Finally, one must avoid the fallacy of contrasting moonlight during the night with light from terrestrial objects during the day; for the only proper comparisons would be that between moonlight and earthlight both during the day (mentioned above), and between moonlight and earthlight during the night (which is impossible to observe on earth).

Given that the earth can reflect sunlight as well as the moon, the rest of the explanation of the moon's secondary light involves various correlations between this lunar light and the light reflected from the earth to the moon (DML 102–3, 113). The chief correlation is that the moon's secondary light is visible when the amount of earth-light reaching the moon is greatest, for the secondary light is visible only when the moon is crescent, and at such a time an observer on the moon would see almost the full disk of the earth lit. More generally, the portion of the terrestrial disk visible from the moon (i.e., the portion reflecting light onto the moon) is the reverse of the portion of the lunar disk visible from the earth: when the moon is thinly crescent, the earth (seen from the moon) is almost full; as the moon waxes, the earth wanes; to a moon in first quarter, there corresponds an earth in last quarter; when the moon is gibbous or almost full, the earth would appear as thinly crescent; and so on. By the time the moon grows to half, its secondary light disappears, both because the earthlight is insufficient to be even faintly visible, and because the primary moonlight (coming directly from the sun) is enough to overwhelm the secondary light.

Another correlation connecting the moon's secondary light with light reflected from the earth is this. Galileo claims to have observed that the secondary light is stronger when the moon is crescent before conjunction (with the sun), as compared to when the moon is crescent after conjunction (DML 112–13). That is, the moon appears crescent when it is near conjunction with the sun; before conjunction, it is seen crescent in the east at dawn before sunrise (and is called a waning crescent); after conjunction, it is seen crescent in the west at dusk after sunset (and is called a waxing crescent). Now, Galileo notes that when the crescent moon appears to him in the east before dawn, the earthlight it receives comes mostly from the Eurasian land mass, whereas when the crescent moon appears in the west after sunset, the earthlight it receives comes largely from the Atlantic Ocean. Add to this that the roughness of a land surface enables it to reflect more light to the moon than the smoothness of an ocean surface, which also implies that to an observer on the moon the terrestrial oceans would appear darker than the continental land masses; these are consequences of the experiments made earlier (IB2). The result is that the moon is receiving less light from the earth when he sees it (waning) crescent in the east than when he sees it (waxing) crescent in the west; it follows that the secondary light in the former configuration will be stronger than in the latter.

All these considerations show that the moon's secondary light can be explained as resulting from sunlight reflected from the earth. This in turn shows that earthlight is a possible explanation of the moon's secondary light. To show that this is the best explanation, Galileo has to show that no other explanation will do the job.

First, there was the traditional explanation that the secondary light is the moon's own inherent light (DML 103–4; cf. Besomi-Helbing 278–79, 308–9). This is conclusively refuted by the fact that during total eclipses of the moon, the secondary light is not visible. For such eclipses occur when the moon moves into the earth's shadow, and hence is receiving no sunlight; thus, there is no overpowering light that would overwhelm the secondary light; and so the moon's secondary light should appear at its strongest then if it were originating from its own body.

A second possible explanation advanced by some, was that the secondary light derived from other heavenly bodies, especially Venus, which is the closest such body to the moon (DML 104–5; cf. Besomi-Helbing 310–11). Against this explanation, the evidence from lunar eclipses again applies. But there is an additional piece of counterevidence: if the secondary light derived from other heavenly bodies, then its strength should increase as the lunar cycle goes from crescent to half-moon; for as this happens in the cycle, the moon becomes visible for longer periods during nighttime, and at such times the interference from sunlight decreases, and the effect of the light from the other heavenly bodies would be more visible. However, as already mentioned, observation reveals that the secondary light disappears as the lunar cycle goes from crescent to half-moon.

A third possible explanation (DML 105–10; cf. Besomi-Helbing 311–13) claimed that the secondary light derived from the sun: the moon's body is not completely opaque, but translucent; so sunlight passes through the lunar body, and although it is thereby weakened in the process, it is not completely extinguished and can come out at the other end with some degree of visibility, however faint. This explanation would imply that the secondary light is brighter at the edges of the lunar disk and weaker in the middle, since sunlight has less distance to go through in the region near the edges, as compared to the full diameter to go through near the middle; but this is not in fact observed, although some have wrongly claimed to have observed it. The explanation would also imply that during solar eclipses, the secondary light should be stronger in the part of the lunar disk directly in front of the sun than in the rest of the disk; but, again, this is an invention by the proponents, and no such phenomenon is in fact observed. Finally, there is no truth in the claim that the lunar body is translucent: telescopic observations can conclusively refute this claim, since they reveal that lunar mountains cast sharp shadows as a result of being completely opaque and show no signs of being translucent.

In sum, the moon's secondary light is best explained as deriving from sunlight reflected by the earth onto the lunar surface, and not as due to the moon's own intrinsic light, or to light from other heavenly bodies (especially Venus), or to sunlight passing through the translucent lunar body. Thus, the moon and the earth share an additional similarity, insofar as they both lack intrinsic luminosity but have the ability of reflecting sunlight onto each other.

1B4. DIFFERENCES BETWEEN EARTH AND MOON (DML 113–21)

The similarities between the earth and moon demonstrated so far undermine the earth–heaven dichotomy and show the fundamental unity of the terrestrial and heavenly regions. However, for Galileo, those similarities need to be qualified by an admission of several differences.

One important difference between the earth and moon is that there is probably no water on the moon. The reason for this is that telescopic observation reveals absolutely no clouds or rain on the moon; instead its surface always exhibits the same features in a perfect tranquillity.

Moreover, it would be wrong to argue that there is water on the moon because common observation with the naked eyed reveals large dark spots visible on the lunar surface, and their dark appearance could be explained by saying that these spots consist of oceans of water, by contrast with the brighter areas, consisting of land. This argument is incorrect because there is no reason to prefer such an explanation of the moon spots to other alternatives: the same effect would result if these darker regions were covered with forests, or with rocks and soil of a naturally darker color. Nor can telescopic observation decide this issue, because although the telescope reveals that the surface of these lunar spots is relatively flat, we have no way of seeing whether this flat surface is water or land.

Another difference between the earth and moon is that on the moon the cycle of night and day lasts about one month. For we know that the moon revolves eastward around the earth once a month, and that it always shows the same side toward the earth. The only way for the moon to always show the same side is for it to rotate eastward on its axis once a month; that is, as the moon revolves in its orbit around the earth in a counterclockwise direction, it must be rotating on its own axis in the same direction just enough to compensate for the orbital displacement. Thus, if we disregard the earth and consider the moon's monthly axial rotation only in relation to the sun, any one point on the lunar surface faces the sun and experiences daylight for about half a month, and it is on the far side from the sun and experience darkness for the other half of the month.

A third difference is that there are no significant seasons on the moon. For seasonal variations in the length of night and day and the amount of sunlight received from the sun derive from the inclination of the equatorial plane (or axis of rotation) relative to the plane of the ecliptic. For the earth, this inclination is 23.5 degrees; for the moon, it is about 5 degrees, deriving primarily from the inclination relative to the ecliptic of the plane of the moon's geocentric orbit (which corresponds to the moon's equatorial plane or axial rotation). Thus, in any one region of the moon, the seasonal variation in the amount of light received from the sun would be roughly 1/5 that of the seasonal variation on the earth.

These three differences, in turn, imply a fourth, crucial difference: there is probably no life on the moon, or at least no life as we know it. For life as we know it requires water, and a form of life without water is almost inconceivable. Furthermore, even if life without water were possible, life on the moon would have to be able to withstand the tremendous changes in temperature between night and day, stemming from the fact that the lunar cycle of night and day is, as mentioned above, one month; thus, during daylight it would be extremely hot, and during nighttime it would be extremely cold. Such night–day extremes would be made worse by the lack of significant seasons; that is, the extreme variation between night and day would not be moderated by the seasons, since any one region would always experience the same basic season. Again, life as we know it, cannot survive in extremely hot or cold environments such as these.

This discussion of differences between the earth and the moon is followed by an epistemological discussion of the power and limitations of the human mind. This philosophical discussion is not an afterthought vis-à-vis the discussion of the earth–moon differences, any more than the latter was an afterthought vis-à-vis the discussion of their similarities. In fact, all these discussions are illustrations of Galileo's epistemological modesty and realistic attitude, as contrasted to both pessimistic skepticism and optimistic dogmatism (cf. Sosio 123 n. 1, 125 n. 1).

To elaborate, on the one hand, Galileo holds that the human mind is limited in important ways. It is not a measure of what can occur in nature, since if one ever tries to understand fully anything, one realizes how little one knows (as the example of Socrates shows). Moreover, the human mind is infinitely inferior to the divine mind, both in the number of propositions known and in the manner of knowing them; for God keeps infinitely many propositions simultaneously in His mind, without having to use the step-by-step reasoning process characteristic of human beings. On the other hand, the human mind has some impressive powers. We can know a few propositions, such as those of pure mathematics, with the same "objective certainty" (DML 118) which God has. Moreover, one cannot but marvel at the various artistic and scientific achievements of humanity, e.g., the alphabet and the musical scale. Our knowledge of the moon is a good example of this limited but real power of the human mind: although there are some similarities between the earth and moon, there are also some differences; although we can think of several conditions that might cause the large dark moon spots, we have no way of determining which one it really is; and although the moon cannot sustain life as we know it, it is not impossible for God to have created a completely different kind of life there.

NOTES

1 Johnstone (1959, 1978); Woods 2004: 111–24; Finocchiaro 2005a: 277–91, 329–39.
2 In this parenthetical reference and the next two (as well as in some earlier and later ones), please note that the line numbers in Galilei 1897 (FAV) enable us to track and document the textual accuracy of my interpretation to a very high degree.
3 Cf. Shea-Davie 159; Aristotle, *On the Heavens*, bk 1, ch. 3, 270a13–22, and bk 1, ch. 4, 270b32–271a33.
4 Cf. also DML 63; Aristotle, *On the Generation of Animals*, bk 3, ch. 10, 760b31.
5 Cf. Aristotle, *On the Heavens*, bk 2, ch. 11, 291b10–23; Ptolemy, *The Almagest*, bk. 1, ch. 4 (1952: 8–9); Galilei, *Cosmografia*, in Galilei 1890–1909, vol. 2: 250–51.

5

DAY II
EARTH'S DAILY AXIAL ROTATION

IIA. SIMPLICITY ARGUMENTS FOR TERRESTRIAL ROTATION (DML 123–55)

At the beginning of the Second Day of the *Dialogue*, there are some lengthy methodological discussions on the role of authority and the nature of rationality (DML 123–32, 148–53), which are best analyzed later when we focus on the book's methodological content and significance. There is also a comprehensive summary of the arguments against terrestrial rotation (DML 144–55), which does not need a separate discussion here, but is better taken into account when these arguments are analyzed later. Finally, there is a third part (DML 132–44) that presents arguments in favor of terrestrial rotation, based on the principle of simplicity, and whose strength is merely probable. This is the focus of our reconstruction now.

Galileo begins the elaboration of the simplicity arguments for terrestrial rotation with a statement of the principle of the relativity of motion. This is the idea that motion exists only in relation to things lacking it, whereas shared motion has no effect on the relationship of things sharing it. Let us abbreviate this proposition by the letter A, and then we will progressively label other key propositions in alphabetical order, unless explicitly labeled otherwise in the text under consideration. Such labels are helpful in keeping track of the thread of this long argument. They are also indispensable in grasping how the whole argument hangs together. In fact, the structure of this argument can be represented by the following tree-root diagram, in which the various propositional labels are interrelated by means of connecting lines that indicate what supports what. This diagram is given here for reference, but may be sidestepped for now, and consulted only as the need arises in the exposition.

In the course of the discussion, Galileo gives a partial justification supporting the principle of relativity in terms of familiar examples of shared and unshared motion. However, he does not stress this justification and regards the principle as relatively uncontroversial. Let us label this support B.

His primary interest is to apply this principle to the problem of explaining apparent diurnal motion; this is the easily observable phenomenon that all heavenly bodies are seen to revolve around the earth from east to west in the period of twenty-four hours. The relativity of motion implies that the apparent westward diurnal motion can be explained in two ways: either all heavenly

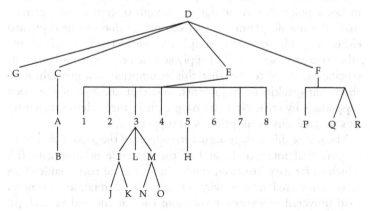

Figure 5.1

bodies actually revolve westward daily around a motionless earth, or the earth alone rotates eastward on its own axis every day. Let C refer to this implied claim.

Given this claim (that diurnal motion can be explained either way), Galileo goes on to argue that therefore (D) terrestrial axial rotation is more likely than universal revolution around a motionless earth, because (E) the geokinetic explanation of apparent diurnal motion is simpler than the geostatic explanation, and (F) nature usually operates by the simplest possible means. The argument so far is: A because B; because A, therefore C; and because C, E, and F, therefore D.

Galileo makes it clear that he is asserting the final conclusion D only with probability, and that his conclusion embodies an implicit comparison between the two contradictory views which generate the controversy. He also explicitly states that his conclusion needs to be qualified in another crucial way; that is, it depends on the assumption that *all other relevant phenomena can also be explained either way*. This assumption, in turn, is equivalent to saying that all the many anti-Copernican objections can be refuted and all the phenomena on which they are based could occur on a moving earth. Let us label the assumption just mentioned proposition G.

The best place for this assumption in the overall structure of this argument is to have it as an additional premise directly support-ing the final conclusion, alongside C, E, and F. Assumption G makes a point that is similar to, though of course more general than, the specific claim (C) that diurnal motion can be explained either way. Moreover, the simplicity considerations made in the other two premises (E and F) apply as much to the general as to the specific point. Note also that this assumption is a generalization about anti-geokinetic arguments, and that the rest of the book supports it by critically examining each in turn. However, within this passage this assumption is unsupported.

The rest of this passage argues in support of the greater simplicity of terrestrial rotation (E) and of the principle of simplicity (F), which so far have just been stated. In its central part, Galileo lists seven numbered reasons why terrestrial axial rotation is simpler than universal geocentric revolution; then at the end he adds an eighth reason, without labeling it as such. That is, the geokinetic system involves (1) fewer moving parts, smaller bodies moving, and lower speeds; (2) only one direction of motion (eastward) rather than two opposite ones (eastward *and* westward); and (3) periods of revolution which follow a uniform pattern, namely that of increasing with the size of the orbit. Moreover, the geostatic system involves (4) a complex pattern for the size and location of the orbits of the fixed stars; (5) complex changes in the individual orbits of fixed stars (due to [H] the precession of the equinoxes); (6) an incredible degree of solidity and strength in the substance of the stellar sphere, which holds fixed stars in their fixed relative positions; (7) a mysterious failure for motion to be transmitted to the earth after it has been transmitted all the way down from the outer reaches of the universe to the moon; and (8) the postulation of an *ad hoc primum mobile*.

In this part of the argument, we thus have eight propositions supporting the comparative simplicity claim E. Although all these are propositions that implicitly compare and contrast the two alternatives in regard to a particular property, only the first two involve matters of degree, whereas the other six involve discrete properties which one of the two world systems possesses but the other one lacks. Moreover, note that the first proposition has three parts, each of which would need separate support; that the fifth one is provided a brief justification; and that the third is especially important and is supported by a relatively lengthy, novel, and strong argument.

Proposition 3 states that the periods of revolution have a uniformity in the Copernican system which they lack in the geostatic one. In the supporting subargument, Galileo begins with a claim which I call the *law of revolution* (I): it is probably a general law of nature that, whenever several bodies revolve around a common center, their periods of revolution become longer as the orbits become larger. He then supports this by the well-known fact that (J) the planets revolve in accordance with this pattern, and by his own discovery that (K) Jupiter's satellites also follow the pattern. The important point is that, although this feature of planetary revolutions was known to the Ptolemaics and incorporated into their system, before the discovery of Jupiter's satellites it would have been rash to generalize a single case into a general law; however, the completely different case of Jupiter's satellites suggested that this was not an accidental coincidence but had general systemic significance. Thus, while this inferential step, "J, K, so I", is not conclusive, it has considerable strength.

Given the law of revolution (I), Galileo proceeds to combine it with the claim that, whereas (L) the earth's diurnal motion in the Copernican system is consistent with the law of revolution, (M) the diurnal motion of the universe in the Ptolemaic system is not. The point is that this difference gives the Copernican system a uniformity lacking in the Ptolemaic system, which is what is asserted by Galileo's third reason (proposition 3) why the former has greater simplicity than the latter (proposition E). He considers the consistency between Copernican diurnal motion and the law of revolution to be sufficiently obvious to need no support. The argument would be that in the Copernican system the diurnal motion is the earth's axial rotation, and this axial rotation is not an orbital revolution and does not involve a member of a series of increasingly larger orbits; thus, the law is not even meant to apply to terrestrial rotation. To justify the inconsistency between the law of revolution and Ptolemaic diurnal motion, he points out that (N) in the geostatic system the diurnal motion corresponds to the revolution of the outermost sphere (whether stellar sphere or *primum mobile*) around the central earth, but (O) this outermost sphere involves both the largest orbit and the shortest period. This completes the subargument supporting the greater Copernican uniformity with regard to periods and orbits of revolution (proposition 3), namely the argument *from* the law of revolution.

At the end of this passage, there is a discussion of the principle of simplicity (F), which is a crucial premise directly supporting the final conclusion about the greater likelihood of terrestrial rotation (D). Although there might be some question about how this principle should be formulated, let us focus on the interpretation adopted above—that nature usually operates by the simplest possible means. When so stated, Galileo in part suggests what might be called a *teleological* justification for simplicity, namely the teleological principle that (P) it is useless to do with more means what can be done with fewer.

However, Galileo also recognizes that (Q) the principle of simplicity is subject to the following *theological* objection: given a God who is all-knowing and all-powerful, God can create and operate a more complicated system as easily as a simpler system, and so a more complex world system is as likely to exist as a simpler one. That is, it is false that nature usually operates by the simplest possible means, because nature was created by an infinitely powerful God, and such a God would be as likely to use more power (or energy) to operate a more complex universe as to use less power to operate a simpler world. Galileo tries to address this objection directly, though I do not find his answer convincing. But let us simply label his answer R, and note that I have labeled by the simple label Q the whole admission of the existence of the theological objection, including a statement of its details as part of proposition Q.

In terms of these labels, this last subargument justifying the principle of simplicity has the structure: "F because P, and because although Q, R."

In sum, the simplicity argument for terrestrial rotation claims that it is more probable that the earth rotates on its own axis daily in an eastward direction than that all the heavenly bodies revolve daily in a westward direction around a motionless central earth (D). The justification is that, given the principle of the relativity of motion (A), the basic phenomenon of apparent diurnal motion can be explained by either the Copernican or Ptolemaic hypothesis (C); but the Copernican explanation is simpler than the Ptolemaic one (E), for at least eight specific reasons (1–8); and nature usually operates by the simplest possible means (F). Nor can one plausibly reject this principle of simplicity (because P, and because although Q, R); nor can one claim that there are other phenomena which can be explained by the Ptolemaic hypothesis but not by the Copernican one (G).

IIB1. GEOSTATIC ARGUMENT FROM VIOLENT MOTION (DML 155–58)

The first of Aristotle's arguments against the earth's motion is paraphrased by Salviati as follows:

> The earth cannot move circularly, because such a motion would be
> violent, and hence not perpetual. The reason why it would be violent
> is that if it were natural, the parts of the earth would also move cir-
> cularly by nature; but this is impossible, because the natural motion
> of the parts is straight downwards.
>
> (FAV 159; cf. DML 155)[1]

This argument is obviously a variant of the geocentric argument
from natural motion, which was discussed earlier (IA4). Whereas
the earlier argument is based primarily on premises about natural
motion, this one is based primarily on premises about violent
motion, which for the Aristotelians is the opposite of natural
motion. Moreover, whereas this argument formulates the conclusion
primarily in geostatic terms, the earlier one formulated its conclusion
primarily in geocentric language.

According to Galileo's criticism, the key flaw in this argument
is ambiguity and equivocation. One ambiguity is embedded in
the talk of the circular motion of the earth's parts. The clause "the
parts of the earth would also move circularly by nature" can mean
either that these parts would naturally rotate around their own
centers, or that they would naturally rotate around the earth's
center. If the former were the meaning, the corresponding claim
would be absurd, namely the claim that if the earth were moving
naturally around its own center, then its parts would also move
naturally around their own (respective) centers. Here the absurdity is palpable, and it is another reminder of the problematic
character of the principle that the same holds for the parts that
holds for the whole. In Galileo's own words, this "would be a
piece of nonsense no less than if one were to say that every part of
the circumference of a circle had to be a circle, or even that since
the earth is spherical every part of the earth must be a ball" (DML
155; cf. FAV 160).

But let us consider the second possible meaning. In that case the
corresponding claim would be that if the earth's circular motion
around its own center were natural, then its parts would be
naturally moving around the earth's center. This claim would be
unobjectionable, but then the problem would shift to the claim that
"this" is impossible, namely that it is impossible that the earth's parts

would move naturally around the center of the earth. The argument grounds this impossibility on the principle of natural straight motion. The difficulties discussed earlier (IA2–3) would now resurface. Specifically applied here, the problem would be the following, which involves an ambiguity about the meaning of "natural."

In asserting that the natural motion of the earth parts is straight downwards, the argument is presupposing a meaning of "natural" such that natural motion is motion spontaneously undertaken by a body when free to move and left undisturbed. This is the sense in which the principle of natural straight motion is true. However, this principle is not true in the sense that heavy bodies move always and eternally downwards, or even in the sense that they could potentially so move, for once they had reached the center their motion would stop. That is, the principle is not true if by natural we mean potentially perpetual. Unfortunately, this other meaning of natural is precisely the notion presupposed by the argument at another point, in the step saying that given that the earth's circular motion were violent or not natural, it would not be perpetual. Thus, the argument is also equivocating with regard to the meaning of natural: it needs the assertion that the earth's circular motion would not be natural in the sense of everlasting, but it justifies this assertion on the basis of a claim about how the earth's parts move naturally in the sense of spontaneously *and* temporarily.

IIB2. TWO-MOTIONS ARGUMENT (DML 158–61)

Aristotle's second argument against the earth's motion is paraphrased by Simplicio as follows:

> except for the *primum mobile*, all the other bodies moving with circular motion seem to fall behind and to move with more than one motion. Because of this, it would be necessary for the earth to move with two motions. If this were so, there would necessarily have to be variations in the fixed stars. But this is not seen; instead, the same stars always rise at the same places and always set at the same places, without any variations.
>
> (FIN 143; cf. DML 144–45)[2]

This argument is extremely puzzling, and Galileo's criticism may be regarded primarily as a clarification. But before his critical clarification, some more preliminaries will be useful.

The proposition with which the argument begins is intended to be essentially an empirical generalization about the motions of the heavenly bodies. Aristotle has in mind the fact that the planets can be seen not only to revolve from east to west around the earth every day, but also to usually lag behind each day by a small amount, as if they were simultaneously slowly revolving in the opposite direction, so that in any one period of twenty-four hours they do not quite complete a full revolution. Thus, each planet was also attributed a geocentric revolution in the opposite direction (from west to east) with a particular period that varied from about one month for the moon and two years for Mars to twenty-nine years for Saturn. It is in this sense that each planet had two circular motions, the westward diurnal motion shared by all, and its own characteristic orbital revolution in the opposite direction at a much slower speed.

Furthermore, in Aristotelian natural philosophy, the *primum mobile* (which means literally the "first body in motion") was a sphere lying outside the celestial sphere and was acted upon by the First Unmoved Mover. By rotating daily, the *primum mobile* carried along all the other spherical layers inside it, namely all the other heavenly bodies (except the earth). The *primum mobile* was needed because the celestial sphere could not be regarded as the source of the diurnal motion; in fact, it was assumed that each distinct movement had to derive from a distinct sphere, and there was observational evidence that the celestial sphere had another slower movement. This evidence was the precession of the equinoxes, which had a period of thousands of years, and was in a direction opposite to the diurnal motion; thus, the circular motion of the celestial sphere was considered to be another instance of the generalization about two motions.

Galileo criticizes this argument in several ways (DML 159–61; cf. Finocchiaro 1980: 380–86). First, the argument is confusing and ambiguous, insofar as it is unclear whether it is trying to refute the earth's rotation around its own axis, or its orbital revolution around the sun, or both. The first premise (the just mentioned empirical generalization) seems to be talking about axial rotations, since the circular motions to which it refers are supposedly rotations (around the center of the earth and of the universe) by the heavenly spheres to which the heavenly bodies are attached. However, the talk of two terrestrial motions would seem to refer to the earth's daily axial rotation and its annual orbital revolution around the sun. Thus, there seems to be some switching of meaning in the crucial term "circular motion," and this is an illegitimate move typical of equivocations.

Second, the first step in the argument infers that if the earth moved circularly it would have to have two motions, on the basis of the generalization about two motions. This inference obviously presupposes that the earth is not itself the *primum mobile*, which is claimed to be an exception to the generalization. But to conceive the earth as the *primum mobile* is equivalent to attributing the daily rotation to the earth. Hence, the first step is assuming that the earth does not move circularly, in the sense of axial rotation; but this is what the argument is trying to prove, at least in one of its versions. So, the argument is circular and begs the question.

Third, the gist of the argument is that the earth does not move circularly because it does not have two motions. This presupposes formally that if the earth lacks two motions then it lacks circular motion, namely that if the earth has circular motion then it has two motions. Here the assumption seems to be that terrestrial circular motion would imply two terrestrial motions. However, to attribute circular motion to the earth is to regard it as the *primum mobile*, which in virtue of Aristotle's generalization (mentioned above) does have only one motion, and so if the earth moves circularly then it lacks two motions. Now the assumption appears to be that terrestrial circular motion would imply the absence of two terrestrial motions. The two assumptions seem to conflict with each other. Thus, there appears to be some kind of self-contradiction or internal incoherence in this argument.

Finally, let us focus on the intermediate step of the argument, which amounts to this: if the earth moved circularly with two motions, then there would be changes in the rising and setting of the fixed stars; but such changes are not observed; therefore, the earth does not move circularly. Galileo thinks that although the conditional claim here is false, there is something importantly right about it. That is, it is not true that if the earth had an annual orbital revolution around the sun besides a daily axial rotation, there would have to be changes in the daily rising and setting of the fixed stars. What is true is that there would follow some kind of change in the apparent position and brightness of the fixed stars. However, Galileo does not elaborate this point in this passage. Instead he refers to a later discussion in the Third Day (IIIB2–3), where this topic will be examined in detail, and he ends the discussion with the following appreciation of Aristotle: "here I wish indeed to excuse Aristotle's error, and even to praise him for having hit upon the most subtle argument against the Copernican position which can be found. And if the objection is an acute and apparently cogent one, you shall see how much more subtle and ingenious is its solution; one not to be discovered by a mind less penetrating than that of Copernicus. From the difficulty of understanding it you will be able to infer how much greater was the difficulty of finding it in the first place" (DML 160).

IIB3. VERTICAL FALL (DML 161–64)

The third argument in Aristotle's sequence is the geocentric argument from natural motion, but Galileo reminds us that it has already been examined in the First Day (IA4). Thus, he goes on to Aristotle's fourth argument, which tries to prove the impossibility of terrestrial rotation based on the observational fact that bodies fall vertically. Galileo formulates the argument as follows:

Aristotle says that a most certain argument for the earth's immobility is based on the fact that we see bodies which have been cast upwards return perpendicularly by the same line to the same place from which they were thrown, and that this happens even when the motion reaches a great height; this could not happen if the earth were moving because, while the projectile moves up and down separated from the earth, the place of ejection would advance a long way toward the east due to the earth's turning, and in falling the projectile would strike the ground that much distance away from the said place. Here we may also include the argument from the cannon ball shot upwards, as well as another one used by Aristotle and Ptolemy, namely that one sees bodies falling from great heights move in a straight line perpendicular to the earth's surface.

(FIN 155–56; cf. DML 161, Aristotle 296b22–26)

Since the perpendicular direction mentioned in this argument is perpendicular to the horizontal, let us speak of "vertical" and call this reasoning the geostatic argument from vertical fall. Let us also reconstruct the key step in this argument as follows: if the earth rotated, then bodies would not fall vertically; but bodies do fall vertically; therefore, the earth does not rotate. This inference is formally valid, being an instance of a conditional argument of the form of denying the consequent; that is, the conclusion is correctly deduced by combining two premises, a conditional proposition and another proposition that denies its consequent ("then") clause. Thus, Galileo does not question the correctness of this inference; instead he begins his critical analysis by asking how one knows the truth of the second premise.

Simplicio answers that this is known from observation: bodies fall vertically because they are seen to fall vertically. Galileo agrees with the observation: it is unquestionable that bodies appear to our eyes to fall vertically. But this justification is assuming that apparent vertical fall implies actual vertical fall. And here is the locus of Galileo's next question: how can one show that apparent vertical fall implies actual vertical fall?

This is a crucial question because, as he goes on to argue, this implication does not hold unless the earth stands still; for on a rotating earth apparent vertical fall would imply that the direction of fall was actually slanted. Therefore, to claim that apparent vertical fall implies actual vertical fall presupposes that the earth stands still. But the proposition that the earth stands still is the conclusion this argument is trying to prove. Therefore, the argument from vertical fall assumes the same thing it tries to prove; that is, it is circular and begs the question.

This criticism depends on the claim that apparent vertical fall does not imply actual vertical fall unless the earth stands still; but this claim is correct, and Simplicio himself admits it in the discussion, as does Sagredo, who goes on to summarize this criticism.

The criticism also presupposes the distinction between actual and apparent vertical fall, but it is not obvious how one could question this distinction; in fact, in the next two speeches Sagredo and Simplicio go on to reformulate the original vertical fall argument in a way immune to this criticism, and so they may be interpreted as granting the correctness of this question-begging charge.

The distinction between apparent and actual vertical fall and the fact that they do not coincide on a rotating earth may be explained further by means of the following figure (Figure 5.2). This is not found in the Galilean text, but corresponds to it, and may be useful.

In this figure, the semicircles on the left side represent portions of the earth's equator looked at from above the north pole, and the structures labeled AB and A′B′ represent towers on the earth's surface. The right parts of the figure are simply highly magnified representations of the situations on the left, so that the earth's surface appears flat because the distance involved (BB′) is very small. In all cases, we are imagining looking at the earth from a hypothetical motionless point in outer space. The top part (a) of the figure represents a motionless earth, whereas the middle part (b) and the bottom part (c) represent the earth undergoing axial rotation from west to east (or clockwise in this view), as suggested by the arrows. In parts (b) and (c) of the figure, each situation exhibits two different positions of the tower: the *unprimed* position (AB) represents the tower's position at the beginning of the experiment of dropping a rock from the top of a tower and letting it fall freely; the *primed* position (A′B′) represents the tower's position at the end of the experiment when the fallen rock has reached the ground. Along the vertical wall of the tower, and between towers, there are solid and dotted lines: solid lines represent apparent vertical fall, namely the trajectory as seen by an observer standing near the tower; and dotted lines represent actual vertical fall, namely what would be happening in reality (or at least from the fixed vantage point of an unmoving extra-terrestrial observer). The lines (whether solid or dotted) between towers are drawn both as straight slanted lines and as parabolic slanted lines; here the main point to note is that they are both slanted, and although the parabolic representation is

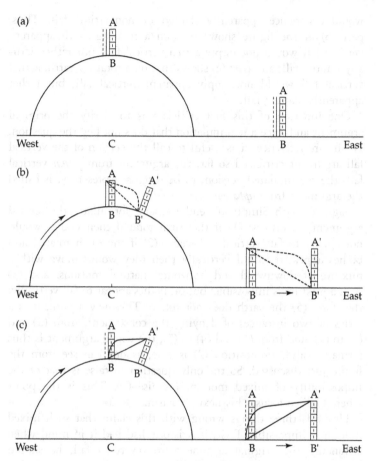

Figure 5.2

more accurate, this refinement plays no role in this discussion. In part (a) of the figure, representing a motionless earth, apparent and actual vertical fall coincide. In part (b), representing a situation where the earth is rotating and apparent vertical fall is experienced on earth, the actual path is slanted or nonvertical. In part (c), representing a situation where the earth is again rotating but where actual vertical fall is taking place, a terrestrial observer would experience apparently slanted or nonvertical fall. Thus, part (b) of the figure shows that on a rotating earth apparent vertical fall would not imply actual vertical fall, but rather actually slanted fall; and part (c) shows that on a rotating earth, actual vertical fall would not imply apparent vertical fall, but rather apparently slanted fall.

One function of this first criticism is to clarify the original argument and force a reformulation that does not beg the question. For future reference, it is useful to call the version of the vertical fall argument criticized so far, the argument from *actual* vertical fall; the reformulated version, to be discussed presently, is called the argument from *apparent* vertical fall.

Sagredo, with Simplicio's endorsement, formulates the revised argument as follows: (1) if the earth rotated then bodies would not appear to fall vertically, because (2) if the earth rotated and bodies appeared to fall vertically then they would move with a mixture of downward and horizontal natural motions, and (3) such a mixture is impossible; but (4) bodies appear to fall vertically; therefore, (5) the earth does not rotate. This new argument is a series of two instances of denying the consequent, from (2) and (3) to (1), and from (1) and (4) to (5); the whole argument is thus formally valid. Proposition (2) is true, as one can see from the figure just discussed. So the only question to raise is that of the impossibility of mixed motion, premise (3). This is the point where Galileo focuses his next criticism.

He finds three things wrong with this claim that such mixed motion is impossible. First, if Aristotle had had it in mind when advancing the original argument from vertical fall, he should have explicitly said so. Here Galileo seems to be suggesting that the argument from apparent vertical fall is an *ad hoc* revision of the argument from actual vertical fall. Even if correct, such a criticism would be more rhetorical than logical.

Second, he claims that such mixed motion is not impossible, which he promises to demonstrate later. I suppose he has in mind the evidence from the ship's mast experiment, just as Simplicio also thinks that the results of this experiment support Aristotle. The explicit discussion of the ship experiment in the next section relates in part to this issue.

Third, Galileo points out that Aristotle himself (*Meteorology* 344a9–14) allows that kind of mixed motion when he admits that fire and the upper atmosphere move upwards by nature and rotate by participation. Simplicio tries to rebut this on the grounds that there is an important difference between the two situations: fire particles are so light that they could easily be carried along by the rotating air, but falling rocks are so heavy that they could not be. However, Salviati could reply that this objection is misconceived because, for Aristotle, fire particles are supposed to be carried along not by the rotating air but by the rotating lunar sphere, which is made of the element aether; and compared to fire particles, this element is even rarer (less dense) than air is compared to rocks. Thus, the third criticism of the revised argument seems cogent; its effect is to make the revised argument seem un-Aristotelian.

In sum, the argument from vertical fall is ambiguous in the sense that the talk of vertical fall can mean either actual vertical fall or apparent vertical fall. When formulated in terms of actual vertical fall, the argument is circular and begs the question, insofar as we cannot know that actual vertical fall is true unless we know that the earth stands still. When formulated in terms of apparent vertical fall, the argument has a problematic premise, namely the claim that a mixture of natural motion toward and around the center of the earth is impossible. For Galileo this claim is false, or at least groundless; in any case, the question of its truth relates in part to the ship's mast experiment, to which the discussion turns next.

IIB4. SHIP'S MAST EXPERIMENT (DML 164–79)

The ship's mast experiment consists of dropping a body, such as a rock or ball, from the top of a ship's mast, both when the ship is motionless and when it is advancing forward. This experiment provided the basis for an interesting and important objection to the Copernican doctrine, specifically to the earth's daily axial rotation. The argument was not explicitly given by Aristotle, but started being developed in the Middle Ages by his commentators, and by the seventeenth century it was a standard part of the

anti-Copernican writings.[3] In Galileo's clear and succinct formulation, the objection reads as follows:

> Because when the ship stands still the rock falls at the foot of the mast,
> and when the ship is in motion it falls away from the foot, therefore,
> inverting, from the rock falling at the foot one infers the ship to be
> standing still, and from its falling away one argues to the ship being
> in motion; but what happens to the ship must likewise happen to the
> terrestrial globe; hence, from the rock falling at the foot of the tower,
> one necessarily infers the immobility of the terrestrial globe.
>
> (FIN 163; cf. DML 167)

This argument has two distinct inferential steps. The first infers
that a body dropped from the top of a tower on the earth's surface
falls to its foot if, and only if, the earth is motionless; and the
adduced evidence is that a body dropped from the top of the mast
on a ship falls to its foot if, and only if, the ship is not moving
forward, assuming that a tower on the earth is like a mast on a
ship. The second step uses the intermediate conclusion inferred in
the first step, to arrive at the further conclusion that the earth is
motionless; that is, given that a body dropped from the top of a
tower falls at its foot if, and only if, the earth is motionless; and
that common observation reveals that a body dropped from the
top of a tower falls at its foot; it follows that the earth is
motionless.

In other words, the first part of the full argument has the form
of an argument from analogy, and its crucial premise is an
observation report from the ship's mast experiment. The second
part is a version of the argument from vertical fall, focusing on
the observation of bodies falling from a tower. Since we have
already examined the vertical fall argument (IIB3), our focus here
is the ship's mast experiment and the first part of the fuller
argument. Galileo criticizes this subargument in two ways: he
undermines its inferential strength by questioning the analogy
(DML 164–66), and he argues that it is false that on a moving
ship the rock falls away from the foot of the mast (DML 166–79).

Galileo argues that the analogy is questionable partly because
on a ship moving forward the horizontal motion imparted by the

ship to the rock before the drop is violent motion, which might be dissipated after the rock is left to itself; whereas on a rotating earth the horizontal motion imparted to the rock before the drop would be natural motion, at least in the sense that it would be everlastingly shared by the rock with the earth, insofar as the rock is a natural part of the whole earth. Another reason for the weakness of the analogy is that on a moving ship air resistance would oppose the horizontal motion acquired by the rock before the drop, since the ambient air would not be moving with the ship, whereas on a rotating earth the air (as a terrestrial body) would probably rotate along; or at least, this would happen for the lower parts of the atmosphere that are trapped between mountains and often contain particles of the elements earth and water. Thus, even if the rock falls behind on a moving ship, it might not do so on a moving earth. However, does the rock fall behind on a moving ship?

Galileo holds that on a moving ship the rock is not left behind but rather falls at the foot of the mast, the same as it does on a motionless ship. He has two reasons for this. One is an experimental report; the other is an indirect theoretical argument. In this passage, he elaborates only the theoretical argument, whereas the experimental report is found only in his "Reply to Ingoli" (1624).[4] However, the argument in this passage is theoretical not in the sense of being a priori, but rather in the sense that its empirical conclusion is based partly on more easily observable phenomena, partly on more easily ascertainable facts, and partly on some theoretical claims that are not arbitrary but can be justified in various ways.

This theoretical argument may be reconstructed as follows. The more easily observable phenomena are that: (1) the undisturbed downward motion of bodies on an inclined plane is accelerated; and (2) their undisturbed motion up an inclined plane is decelerated. The more easily ascertainable fact is that (3) the cause of projectile motion is not the motion of the surrounding air. The reasons for this are as follows (DML 173–79):[5] (3a) wind moves cotton more easily than rocks, and yet rocks can be thrown farther and more easily than cotton; (3b) lead pendulums oscillate much longer than cotton ones; (3c) if the force cannot be impressed directly by the thrower to the projectile, then it cannot be impressed directly by the thrower to the air; (3d) arrows can be shot against the wind; and (3e) arrows travel much less when shot sideways than when shot normally.

From (1) and (2) one may infer that (4) the motion of bodies on an horizontal plane is conserved if undisturbed, and consequently that (5) the horizontal motion which the rock has before being dropped on the moving ship continues even after being dropped, if undisturbed. Now from (3) one can infer that (6) the cause of the motion of projectiles is the impulse conveyed to them by the projector, and consequently that (7) the cause of the horizontal motion of the rock, after it has been dropped, is the horizontal impulse given to it by the hand holding it before dropping. But, (8) there is no way in which this horizontal impulse could be disturbed by the vertically downward tendency due to the weight, because (9) the two are not opposed, but are at right angles to each other, and (10) they have distinct causes (the projector and gravity, respectively). It follows that (11) the horizontal motion of the dropped falling rock is undisturbed, and hence that (12) that motion will continue, and therefore that (13) the rock will end up at the foot of the mast on the moving ship.

This argument is ingenious and plausible, although not completely compelling. It presupposes the principle of the superposition of motions, which is lurking around in the justification of proposition (8) by (9) and (10), and which needs more elaboration; and it presupposes the principle of conservation of motion, a version of which is stated in proposition (4). So it is not surprising that Galileo would seek a direct experimental test of the claim that on a moving ship the dropped rock still falls to the foot of the mast, thus refuting the key premise of the anti-Copernican argument based on the ship's mast experiment. Nor it is surprising that Galileo goes on to elaborate and justify these two principles in the next section.

In sum, the argument based on the ship's mast experiment is a distinct objection to Copernicanism and should not be confused with the argument from vertical fall. The two arguments are related, and their primary relationship is this: the ship's mast experiment attempts to provide evidence supporting a key premise of the argument from apparent vertical fall, the claim that on a rotating earth bodies would not appear to fall vertically, but in a slanted direction. Galileo criticizes the argument from the ship's mast experiment in two ways: first, the analogy between a ship moving forward and a rotating earth is weak, and so even if the falling rock were left behind on a moving ship the same thing would not be likely to happen on the earth; and second, on a moving ship the falling rock is not left behind, but still falls to the foot of the mast, as on a motionless ship.

These two parts of the criticism are complementary and not incompatible. The criticism is cogent, although it depends on the principles of conservation and composition of motion, which were relatively novel and needed elaboration. Finally, it should be clear that such criticism by itself is not an argument for the earth's motion, let alone a proof (cf. Machamer 1973: 27–28); the criticism merely shows that this particular anti-Copernican argument is incorrect, namely that the earth's motion cannot be disproved in this particular manner. This caveat should be as clear to us as it is to Galileo, who at the end of this criticism has the following exchange between Simplicio and Salviati:

> Simplicio. ... so far I do not see that the motion of the earth has been proved.
>
> Salviati. Nor have I pretended to prove it, but rather merely to show why we cannot derive anything from the experiment advanced by the opponents to argue in favor of its standing still, as I believe I can show regarding other experiments.
>
> (FAV 180; cf. DML 179)

IIB5. CONSERVATION AND COMPOSITION OF MOTION (DML 179–95)

The Galilean principle of conservation of motion asserts that horizontal motion, once acquired, is conserved unless it is subject to disturbances or interferences. The principle of the composition (or superposition) of motion states that different motions can be combined in various ways, rather than one totally prevailing over the other: when the components are at right angles to one another, they do not disturb, or interfere with, each other, but each continues independently of the other; when they are in the same direction they add up; when they are in opposite directions they subtract; and when they are along other directions one draws a parallelogram whose sides lie along these directions and whose lengths equal respectively the magnitudes of the two speeds along these directions, and then the body's actual trajectory is along the diagonal of this parallelogram, and its speed is measured by the length of this diagonal. These two principles have already been used, and to some extent elucidated and justified, in the discussion of the ship's mast experiment above. However, much more needs to be said by way of elaboration.

Another example is the following, described in the text as remarkable, novel, and very beautiful. Suppose one shoots a gun in an horizontal direction from the top of a tower so as to hit targets on the ground at different distances. Consider now the motion of these bullets that travel distances of, for example, 1,000 cubits, or 4,000, or 10,000. Let us ask now how long these projectiles will take to hit the target. The answer is that they will all take the same amount of time, and that this is also equal to the time required for one of them to fall freely to the foot of the tower if simply dropped from the top of the tower. Such isochronism is the consequence of the principle of composition of motion, which tells us that after the projectile is fired from the gun, it undergoes a downward motion of fall (due to gravity) that is not disturbed by the horizontal motion which it receives from the firing, and which also continues without interference from the downward motion.

Then there is the example of a horse rider dropping a ball while the horse is galloping. After the ball reaches the ground, it can be seen to move forward in the same direction as the horse, until the interference of the ground features makes it stop. This will happen even if the rider, instead of simply dropping the ball, throws it backwards, as long as such backward speed is less than that of the horse. On the other hand, when these two speeds are equal, the ball will stop when it reaches the ground; and when the rider throws the ball backwards with such force as to give it a speed greater than the forward motion of the horse, then the ball can be seen to move backwards after it hits the ground. All these observational facts are consequences of the principle of conservation of motion.

Next, consider what can be observed in various games: in the game played with those wooden disks that can be thrown in various ways while being given a spin by means of strings wound around their rim; in the game of tennis as a result of the player using the racquet to give the ball both a forward motion and a spin; and in games of bowling, such as *bocce* ball, where again a player can give the ball both forward motion and spin. Such projectiles sometimes move faster after they hit the ground than when they were in mid air; what is happening then is that the spin does not affect the forward motion as long as the projectile is in the air, but does so as soon as it hits the ground; in particular, if the spin is appropriately in the same direction as the forward motion, then the spinning motion is added to the forward motion and the speed of the latter increases. Another observation is that when such projectiles bounce on and off the ground as they move forward, sometimes it happens that their forward motion is faster over a later portion than over an earlier segment of the whole path; what is happening here is that on falling to the ground, such a projectile may hit a rock slanting in the direction of the forward motion, and so it may acquire some spinning, which is then added to the forward motion. Thus, from the principle of conservation one can understand the apparently paradoxical phenomenon that the motion of some projectiles can increase, when a basic part of their motion is conserved and other amounts are added by various secondary factors.

Finally, let us consider the problem of determining the actual path of a freely falling body on a rotating earth. Here we can think of the body being dropped from the top of a tower, and what we are seeking is the shape of the trajectory between that point and the point where it would land on the ground. To solve this problem we need to apply the conservation of motion in the sense that after the body is dropped it would retain the horizontal eastward motion it would have due to the earth's rotation; and we need to apply the composition of motion in the sense that we need to combine properly this horizontal motion with the downward motion due to gravity. From the ship's mast experiment, we can predict that the landing point on the ground would be the foot of the tower. Let us also assume that such a trajectory, if continued unimpeded below the ground, would eventually reach the center of the earth. If the downward speed of fall were constant, then the path would be a spiral, as Archimedes showed. But we know that the falling motion is accelerated, and so to solve our problem we would need the precise formula by which the speed of falling bodies increases; this formula will be stated in the passage on falling to earth from the moon (IIC1), but is not taken into account here.

Instead, here Galileo discussed the possibility that the path in question is an arc of the semicircle whose diameter is the line from the drop point at the top of the tower to the earth's center. He extols the elegance of this idea: for it means that on a rotating earth, whether or not the body is dropped or held at the top of the tower, its absolute motion would be (1) circular, (2) covering equal distances, and (3) uniform in speed. However, he himself calls this idea a *bizzarria*, which literally means "something bizarre," and which has been variously translated as: furie (meaning, inspired exaltation), fantastical conjecture, curiosity, and fancy.[6] Moreover, he regards it as merely probable (DML 191–92). Furthermore, he has Sagredo endorse it with the qualification that "I cannot believe that the center of gravity of the falling body describes any other line than one similar to it" (FAV 192; cf. DML 192, Fréreux-Gandt 186). Finally, he has Salviati end the discussion with the caveat that here we are dealing with an approximation: "I don't want to state categorically for now that the motion of falling objects occurs in exactly this way; but I will say that if the line followed by a falling object is not exactly as I have described it, it is very close to it" (Shea-Davie 250; cf. DML 194).

Nevertheless, from the point of view of the principles of conservation and composition of motion, the potential relevance and illustrative power of this problem remain.

IIB6. EAST-WEST GUNSHOTS (DML 195–98)

The last three anti-Copernican arguments (from actual vertical fall, from apparent vertical fall, and from the ship's mast experiment) attempted to prove the earth's rest based on the behavior of freely falling bodies, although we have seen that the issue of the direction of free fall becomes intimately tied to the issue of the path of projectiles (simply because on a rotating earth freely falling bodies would be projectiles). Another group of objections exploited more explicitly the behavior (real or alleged) of projectiles; it had become increasingly popular after the invention of firearms, and in the sixteenth century Tycho was a leading proponent of such gunshot arguments.

One of these arguments involved comparing and contrasting the range of gunshots in the same direction (eastward) as terrestrial rotation, as well as against it (westward). Galileo formulates the argument as follows:

> [On a rotating earth] if one shoots a cannon aimed at a great elevation toward the east, and then another with the same charge and the same elevation toward the west, the westward shot would range much farther than the eastward one. For when the ball goes westward and the cannon (carried by the earth) goes eastward, the ball would strike the ground at a distance from the cannon equal to the sum of the two journeys (the westward one made by itself and the eastward one of the cannon carried by the earth); by contrast, from the journey made by the ball shot toward the east, one would have to subtract the one made by the cannon while following it. For example, given that the ball's journey in itself is five miles and that at that particular latitude the earth moves forward three miles during the ball's flight, in the westward shot the ball would strike the ground eight miles from the cannon (namely, its own westward five plus the cannon's eastward three), whereas the eastward shot would range two miles (which is the difference between the five of the shot and the three of the cannon's motion in the same direction). However, experience shows that the ranges are equal. Therefore, the cannon is motionless, and consequently so is the earth.
>
> (FIN 146; cf. DML 146–47, Besomi-Helbing 400–401)

In his criticism, Galileo agrees that observation shows that the range of eastward gunshots is the same as the range of westward ones, but he questions the claim that on a rotating earth westward gunshots would reach farther (from the gun) than eastward ones. His key critical conclusion is that on a rotating earth gunshots toward the east and toward the west would range equally, as both sides agree that they would on a motionless earth and as they are observed to do in reality.

The first part of his critical argument is that if on a rotating earth westward gunshots ranged farther than eastward ones, then arrows shot from a crossbow mounted on a moving cart would range farther when shot backwards than when shot forward; but this is not the case, as both experiment and reasoning can show. He gives no experimental report, but he does elaborate the details of the reasoning. Before reconstructing such details, let us, however, clarify the analogy implicit in this part of the argument.

Here Galileo is making an analogy between a rotating earth and a cart pulled by fast moving horses over flat ground. But in his critique of the anti-Copernican argument from the ship's mast experiment (IIB4) he criticized the analogy between moving ship and rotating earth as weak. Is he then being inconsistent? I do not think so, for there is a difference between the two situations; the two analogies are not themselves analogous! For in the ship–earth comparison, the disanalogy applied to the attempt to extrapolate from the falling rock being left behind on a moving ship to the same thing happening on a moving earth; and it applied because there were possible causes for the effect on the moving ship that were not present on the rotating earth, namely the resistance of the air and the non-natural horizontal motion of the rock. However, these same considerations would not apply to the attempt to extrapolate from the rock falling to the foot of the mast to its falling to the foot of the tower; and so Galileo holds that once we observe that the rock falls to the foot of the mast, we can conclude that on a rotating earth the rock would fall to the foot of the tower. Now, the comparison between arrows reaching equally from a moving cart and gunshots ranging equally on a rotating earth reflects the latter type of situation where the same mechanisms occur and there are no causes for differences to emerge.

But how does Galileo argue that when we mount a crossbow on a cart and shoot arrows in opposite directions when the cart is in motion, the arrows shot forward reach as far (from the cart) as the arrows shot backwards. His argument is a *reductio ad absurdum*;

that is, he assumes the opposite (i.e., that the anti-Copernicans are right) and then he argues that this leads to a contradiction.

Suppose then that arrows shot backwards from a moving cart range farther than those shot forward, where by range is meant the distance between the cart and the arrow at the time it hits the ground. Even if this were the case, there would still be a way of arranging for the backward and forward arrows from a moving cart to range equally: all we would have to do is to shoot the arrows forward with greater force and backwards with less force, by adjusting the tension in the crossbow's spring that gives the arrows their propulsive power. For example, if with a given spring tension the arrows shot from a motionless cart reach 300 cubits, and if on a moving cart the same spring tension sends the forward arrow 200 cubits and the backward arrow 400 cubits, then on the moving cart we could increase the spring tension by some calibrated amount (say by one third) when shooting the arrow forward and decrease it by the same amount when shooting the arrow backward.

The next point to understand is that the various spring tensions achieve the various results by imparting various speeds to the arrow. For example, when we shoot from a motionless cart using a spring tension that sends the arrow to a distance of 300 cubits, we may take the arrow's speed to be three units. When this tension is increased by a third, the speed conveyed by the crossbow to the arrow would be 4 units; and when the tension is decreased by a third, the speed conveyed by the crossbow would be 2 units. Now, from a motionless cart, these changed speeds would take the arrow to distances of, respectively, 400 cubits and 200 cubits.

Moreover, from a moving cart, the speed acquired by the arrow depends not only on the spring tension but also on the motion of the cart: when we are shooting forward, the speed acquired by the arrow is increased compared with the speed conveyed by the spring alone; whereas when we are shooting backward the arrow's speed is decreased. Thus, at constant spring tension, the arrow shot forward from a moving cart has a faster speed than the arrow shot from a motionless cart, and so it will travel a correspondingly greater distance, say 400 cubits; but in the meantime the cart is moving in the same direction, say 100 cubits; thus the arrow from the moving cart reaches the ground at a distance of 300 cubits from the cart. On the other hand, the arrow shot backward from a moving cart has a slower speed than the arrow shot from a motionless cart, and so it will travel a correspondingly shorter distance, say 200 cubits; but in the meantime the cart is moving in the opposite direction, for example by 100 cubits; and so the arrow from the moving cart also reaches the ground at a distance of 300 cubits.

In short, from the assumption that arrows shot backward from a moving cart hit the ground at a greater distance from the cart than arrows shot forward, we have derived the consequence that they both land at the same distance. Therefore, this assumption must be false. It is not true that arrows shot forward and backward from a moving cart range differently; instead they range equally, the same as when shot from a motionless cart.

This indirect argument can be reformulated more directly by reasoning explicitly in terms of the situation of gunshots on a rotating earth, as suggested by Galileo when he says: "now apply this reasoning to the case of artillery, and you will find that, whether the earth moves or stands still, gunshots of the same power will range equally, regardless of the direction they are aimed at" (FAV 196; cf. DML 198). The direct argument can be reconstructed as follows.

If the earth rotates, then the eastward projectile is moving at a greater absolute[7] speed than the westward projectile, since the absolute speed of the eastward projectile is the sum of what it receives from the gunpowder and from the moving earth, whereas the absolute speed of the westward projectile is the difference between these two same quantities. But if the eastward projectile moves absolutely faster, then it travels a greater absolute distance than the westward one. If so, then the distance (relative to the moving earth) traveled by the eastward projectile is the same as that of the westward projectile, since for the eastward projectile the relative distance equals its absolute distance less the distance traveled by the gun, and for the westward projectile the relative distance equals its absolute distance plus the distance traveled by the gun. If these relative distances are the same, then the ranges of the gunshots are also the same, since on a rotating earth the observed range of a shot would be the absolute distance traveled by the projectile modified (plus or minus) by that traveled by the gun. Therefore, if the earth rotates, the range of eastward gunshots and that of westward ones would be equal. It follows that east–west gunshots have equal ranges regardless of whether the earth rotates or stands still.

Details aside, the key error of this anti-Copernican argument may be seen as follows. From the hypothesis that the earth rotates it tries to deduce a consequence about projectiles, which is not in fact observed. This consequence is deduced by adding up some relevant distances (the westward trajectory of the projectile and the eastward distance traveled by the gun on a rotating earth), and by subtracting some others (the eastward distance traveled by the gun from the eastward trajectory of the projectile). But in so doing it forgets to take into account that on a rotating earth the relevant speeds are also different, smaller in the former case and greater in the latter case, just enough to make up for the different distances. It's as if, in trying to examine the consequences of the earth's rotation, the anti-Copernicans forget to take into account the earth's rotation. In Galileo's own eloquent words, "the error which Aristotle, Ptolemy, Tycho, you, and everyone else have made, is based on the fixed and deep-seated impression that the Earth is at rest, which you are incapable of shedding even when you want to speculate about what would happen if the Earth were in motion" (Shea-Davie 254; cf. DML 198).

IIB7. VERTICAL GUNSHOTS (DML 198–206)

One of the many anti-Copernican arguments based on the behavior of gunshots involved shooting in a vertically upward direction and observing the shell fall back to the gun. Salviati formulates it as follows:

> We should go on now to speak of ordnance mounted perpendicular to the horizon, that is of a shot towards the zenith, and finally of the return of the ball by the same line onto the same piece, although in the long time in which it is separated from the piece the Earth has

transported it many miles towards the east. Now it would seem that the ball ought to fall a like distance from the piece towards the west. This does not happen; therefore, they say, the piece did not move but stayed and waited for the ball.

(Santillana 188; cf. DML 202)

This objection is a version of the argument from vertical fall, but it possesses a special strength. Such strength lies in the fact that the process mentioned here takes much longer than the experiments appealed to in the other versions: dropping a rock from the top of a tower; dropping it from the top of a ship's mast; throwing it vertically upwards with one's hands and having it return there; or using a bow to shoot an arrow vertically upwards. The longer duration and greater height involved in the vertical gunshot would magnify the presumed westward deviation, since the gun carried by the earth would be moving a longer distance eastward.

Despite this special strength, this objection can be refuted analogously to the way the other versions of the vertical-fall argument were refuted. Galileo explicitly says so (Santillana 188, DML 202), without repeating any of the earlier criticisms. However, he exploits the present version to discuss some relatively novel points, which he had not raised in the earlier critiques.

The main new issue concerns the claim made in the second sentence of the passage just quoted: that if the earth rotated then the projectile shot vertically would fall a long distance west of the gun (rather than back to the gun). This claim is analogous to an assertion that is part of the argument from apparent vertical fall: that if the earth rotated then bodies could not *appear* to fall vertically. Earlier, this assertion was justified by saying that if the earth rotated and bodies appeared to fall vertically, in reality during their fall they would be moving with mixed motion (toward and around the center); but such mixed motion was impossible for the Aristotelians. The issue then became whether such mixed motion is possible. However, in the present context Simplicio tries a different justification of the crucial claim.

The justification now is that if the earth rotated and bodies appeared to fall vertically, in reality bodies would be undergoing slanted fall; but if that happened, our senses would be revealing apparent vertical fall while in reality actually slanted fall would be occurring; this in turn would mean that our senses would be deceiving us; and such a deception of the senses is absurd, because they are our basic instrument for learning about the world. Applied to vertical gunshots, this subargument would be: if the earth rotated then a projectile shot vertically would not fall back to the gun, because if that happened then in reality the projectile would be moving in a transversal direction, while to our senses it would appear to be moving vertically; that is, we would not be seeing what was really happening, but rather would be experiencing a deception of the senses, which is absurd.

Galileo's critique has two main elements. One is an incisive diagram showing that on a rotating earth a vertical gunshot would embody the experience of seeing an actually slanted trajectory as an apparently straight perpendicular one. The other element is an elaboration of the principle of the relativity of motion to reply to Simplicio's difficulty about the deception of the senses.

The diagram is the following:

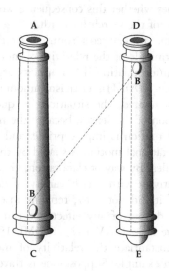

Figure 5.3

AC is a cannon aimed vertically upwards, and B is the ball shot in that direction. Consider just the path of the ball inside the barrel of the gun, after the cannon is fired. On a rotating earth, while the ball moves along the barrel, the cannon is carried by the earth's rotation and moves from position AC to DE. This means that the actual path of the ball would be a transversal line like BB, going from the bottom of the first position of the gun, C, to the top of the second position, D. However, a terrestrial observer looking at this process would also be carried by the earth's rotation from along the position AC to DE, and so would only see the ball being fired from the bottom of the gun and exiting at the top, having moved along the barrel, without any transversal deviation.

Galileo is aware that the amount of slant would depend on the eastward speed of the gun due to terrestrial rotation, as compared to the speed imparted by the gun to the cannon ball, and that the precise shape of the trajectory would depend on how the ball's speed is slowed by gravity. But these complications are not elaborated here, where the main point is to illustrate that, indeed, on a rotating earth our senses would not detect some motions. The issue now becomes whether this consequence would be devastating for the enterprise of the search for truth.

Thus, we come to Galileo's solution to this difficulty. He appeals to the principle of the relativity of motion, which he also utilized earlier in formulating the simplicity argument in favor of terrestrial rotation (IIA). His criticism amounts to pointing out that if the earth rotated, the situations in questions would not involve a deception of the senses, because the nature of motion is such that shared motion is imperceptible, and in these situations the horizontal rotational motion is shared by the observer and the observed projectiles. By way of elaboration, Galileo formulates the principle of relativity with a rich variety of descriptions: shared motion "is as if it did not exist, remains insensible and imperceptible, and is devoid of any effect; we can observe only that motion which we lack" (FAV 197; cf. DML 199).

To further substantiate the relativity of motion, Galileo has Sagredo relate an example. Suppose one is traveling by ship from Venice to Syria. At the beginning of the trip an artist starts drawing on paper a picture with people, animals, and some scenery; he works on the painting on and off for the whole trip, and finishes it when they arrive at destination. Now the actual path of the tip of the artist's pen is a line some thousands of miles long with countless very minor irregularities due to the various rocking motions of the ship, and more importantly due also to the motion of the artist's hand. Obviously, the only motion of the pen that gets recorded on paper and generates the picture is the motion deriving from the artist's hand, which is not shared by the paper; the long, forward motion of the ship, and its countless swaying motions, are without effect on the drawing paper, as if they did not exist. There is no deception here in what we experience by observing the paper; there is no failure on our part to see anything we should be seeing since the rest of the pen's motion is not anything susceptible of being seen; such is the nature of motion.

The case of observing moving bodies on a rotating earth is similar. We are not being deceived when we see the vertical gunshot move vertically, and do not see it move transversally; for the latter motion is not there to be seen for us standing on the earth and carried by its rotation. Nor are we disregarding sense experience as a key instrument to learn about the world.

IIB8. NORTH-SOUTH GUNSHOTS (DML 206–8)

Besides shooting toward the east and west and toward the zenith, another direction which the anti-Copernicans believed creates difficulties is north or south. That is, they believed that on a rotating earth, gunshots along the meridian toward the north or toward the south would deviate westward and miss their target. Galileo states the argument as follows:

> shooting toward the south or toward the north also confirms the earth's stability. For one would never hit the mark aimed at, but instead the shots would always be off toward the west, due to the eastward motion of the target (carried by the earth) while the ball is in midair.

> (FIN 146; cf. DML 147, Besomi-Helbing 401)

This difficulty also raises issues analogous to those of some of the previous ones, and could be resolved in part by adapting some of the previous answers. However, this brief passage does not do that; instead it discusses a new aspect of the situation.

Galileo has a two-fold criticism of the argument from north-south gunshots. In the first part, he makes an analogy between shooting north or south on a rotating earth and the practice of hunters shooting at flying birds. The explanation of how and why hunters can hit a bird in flight starts with the following basic consideration: when they aim at such a moving target they move the rifle in such a way as to follow the flying bird with their line of sight; this means that the bullet leaving the barrel receives not only a component of motion along the barrel due to the gunpowder, but also a component parallel to the bird's motion due to the motion of the rifle during the process of aiming and firing; thus the bullet not only travels the distance between the hunter and the bird, but also to some extent follows the motion of bird, so as to hit the bird at the location where it has moved during the time the bullet takes in its trajectory. The suggestion is that on a rotating earth, gunners shooting at a northward or southward target could hit it accurately in the same way hunters can hit flying birds; for on a rotating earth, carried by the earth's motion the gun aiming at a target would be itself moving and following the motion of the moving target, and so the bullet would not only receive forward motion from the gunpowder, but also lateral eastward motion from the earth's rotation.

The second part of Galileo's criticism exposes the weakness of this analogy, and elaborates an important clarification. First, in the hunter's shooting, the lateral motion imparted by the rifle to the bullet before firing is much smaller than that which the bird has, and thus it would not be sufficient to enable the bullet to keep up with the bird, and so the bullet would only approach the bird at some point behind it and miss the target. In fact, there are some additional factors which explain how hunters are able to hit flying birds: to a small extent, they anticipate the bird's position and aim at a point ahead of where it is while they aim; furthermore, instead of using a single large bullet, they use a large number of small pellets, which spread during flight and can hit a larger area of the target; and these projectiles move at a very high speed, and so they reach the bird before it has had the time to move much. Thus, the explanation of how hunters hit flying birds is more complicated than the simple one mentioned above.

However, in the case of shooting north or south on a rotating earth, the target aimed at and the aiming gun would be moving at speeds that are approximately equal; thus, the projectile would receive a lateral eastward motion that would enable it to keep up with the target, and so gunners could hit accurately without the complications required for the case of bird hunters. Therefore, there is an important difference between the two cases. However, the resulting disanalogy does not undermine Galileo's criticism of the objection from north-south gunshots, since that difference works against this objection and in favor of the criticism; for the key point is that on a rotating earth, north or south gunshots would not swerve westward by a significant amount and thus miss the target.

On the other hand, there is an important clarification that needs to be made explicit in this discussion. We have seen that Galileo argues that the simpler explanation of how hunters aim at birds is incomplete because the lateral speed imparted to the bullet by the motion of the rifle is much smaller than that which the bird has. Obviously the two relevant speeds are linear, rather than angular, speeds. It is true that the hunter follows the motion of the bird by rotating the rifle, and for this to happen the angular speed experienced by the bullet during this process must equal the apparent angular speed of the bird as observed by the hunter. However, the point is that these two equal angular speeds correspond to very different circles, and so to very different linear speeds. And Galileo makes it clear that it is linear speed that matters in such cases.

This clarification emerges in another way in this passage. When Galileo asserts that in the case of a rotating earth the speed of the target is the same as that of the gun, he adds this qualification: "although the cannon will sometimes be placed closer to the pole than the target and its motion will consequently be somewhat the slower, being made along a smaller circle, this difference is insensible because of the small distance from the cannon to the mark" (DML 208). That is, as I phrased it earlier, the two speeds are only approximately equal. Galileo is saying that although on a rotating earth the gun and the target are moving at the same angular speed, the corresponding linear speeds are different because they are on a spherical surface and located at different latitudes. However, in the present case, involving just a few miles in distance, the difference in linear speeds is insignificant. And this suggests, again, that what is important is linear, not angular, speed.

Finally, there is another important idea suggested by both the criticism of the hunter–artillery analogy and the qualification about significant vs. insignificant differences of speeds. That is, presumably, if and when the distance between the gun and the target (their difference in latitude) is great, the projectile's deviation might be significant. This seems to me to be an intuition of the coriolis effect of modern physics: this is a phenomenon observable in the behavior of terrestrial bodies that move long distances over the earth's surface, such as winds, ocean currents, and artillery projectiles; it is the tendency of these bodies to swerve sideways, rightward in the northern hemisphere and leftward in the southern hemisphere, as a result of the earth's rotation. In the northern hemisphere, a projectile moving southward (the example in Galileo's qualification) has a lower lateral speed than the target, and so it lags behind and is deflected *westward*, which also equals *rightward* (for an observer traveling along with the projectile); and a projectile moving northward has a higher lateral speed than the target, and so it advances forward and is deflected *eastward*, which still equals *rightward*. In the southern hemisphere, it is the projectile moving *southward* that has a higher lateral speed than the target, and advances *forward*, and is deflected *eastward*, which equals *leftward*; whereas it is the projectile moving *northward* that has a lower lateral speed than the target, and lags behind, and is deflected *westward*, which still equals *leftward*.[8]

IIB9. POINT-BLANK GUNSHOTS (DML 208–12)

The last of the gunshot arguments against the earth's rotation involved the behavior of projectiles shot point-blank, that is, in a horizontal direction without pointing the gun at an angle. The best statement of this objection is the following:

These shots along the meridians would not be the only ones that would hit off the mark. If one were shooting point-blank, the eastward shots would strike high and the westward ones low. For in such shooting, the ball's journey is made along the tangent, namely, along a line parallel to the horizon; moreover, if the diurnal motion should belong to the earth, the eastern horizon would always be falling and the western one rising (which is why the eastern stars appear to rise and the western ones to fall); therefore, the eastern target would drop below the shot and so the shot would strike high, while the rising of the western target would make the westward shot hit low. Thus, one could never shoot straight in any direction; but, because experience shows otherwise, one is forced to say that the earth stands still.

(FIN 146–47; cf. DML 147–48)

This argument should not be confused with the east-west gunshot objection.[9] Both arguments do involve shooting alternately toward the east and the west, and both raise the difficulty that on a rotating earth the targets would not be hit properly. However, the east-west argument alleges that when a gun on the ground is aimed high in order to hit a distant target also on the ground, what would be hit instead would be a point that is either farther or nearer than that target; whereas the point-blank argument alleges that when a gun on a tower or hill is aimed horizontally in order to hit a target at the same height on another tower or hill, what would get hit instead would be a point either above or below the target.

Moreover, like all the other gunshot arguments, this one has a crucial step that has the form of denying the consequent: "if p then q; not-q; therefore, not-p." All these arguments have a conditional premise claiming various alleged effects of the earth's rotation, and for Galileo all these conditional claims are problematic. Thus, in the present case, he mentions in passing that, for reasons analogous to those in previous critiques, it is probably not true that if the earth rotated, point-blank gunshots would hit high or low. However, the more important point on which he focuses is to try to determine the amount of deviation which point-blank gunshots would exhibit on a rotating earth. He tries to show that the amount would be so small as to be undetectable, or very difficult to detect. Thus, we are in no position to say that point-blank gunshots do not hit high or low, i.e., to assert the minor premise of the argument step that denies the consequent. We really do not know whether or not this premise is true. Galileo's quantitative reasoning is as follows.

First, consider Figure 5.4, which is very helpful, although it is not found in the text. The arc is a portion of the earth's equator, as viewed from a fixed point above the north pole; O is the point where the gun is fired; WE is tangent to the arc at O and represents the horizon; T_E is a target east of O; T_W is a target to the west. A key assumption is that the projectile fired by the gun at O moves along the tangent WE on both a rotating and a stationary earth. On a rotating earth, T_E' is the later position of the eastern target when the projectile reaches it; T_W' is the later position of the western target; by the time the projectile has reached either target, the horizon is $W'E'$, which can be conceived as the result of rotating the original horizon WE by a certain angle around O, as a function of the earth's rotation.

Next, Galileo goes through the following steps:

1 He estimates that to reach the target the projectile takes about as long as it takes a pedestrian to walk two steps, which would be one second.
2 He reasons that in one second of time, the western horizon rises about 15 seconds of arc, since it turns about 360 degrees in 24 hours; that is, angle WOW' is 15 seconds. Therefore, T_W rises 15 seconds of arc in one second of time, and moves to T_W'.
3 He assumes the distance from the cannon to the target is 500 cubits, which means that OT_W is 500 cubits; thus, the 15 seconds of arc are measured along a circle with a radius of 500 cubits. Then, he disregards the eastward component of the

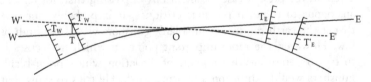

Figure 5.4

target's motion, presumably because this argument is considering only deviations above and below the tangent; thus, OT_W' is also 500 cubits.

4 What Galileo seems to do next is to find the distance from T_W' to the line OW. He approximates this distance by the arc TT_W', whose center is O. In turn this arc is approximated by its chord. These approximations appear to be correct since angle TOT_W' is very small (15 seconds).

5 Using mathematical tables printed in Copernicus's book *On the Revolutions*, Galileo finds that: (a) the chord for one minute of arc, on a radius of 100,000 units, is 30 units; thus, it follows that (b) the chord for one second, on a radius of 100,000 units is 1/2 unit (= [1/60]30); (c) hence, the chord for one second, on a radius of 200,000 units is 1 unit; (d) so, the chord for 15 seconds, on a radius of 200,000, is 15 units; (e) therefore, the chord for 15 seconds, on a radius of 500 cubits, is (500/200,000)(15) cubits = 15/400 cubits = approximately 4/100, or 1/25 cubits, or about one inch.

6 Consequently, the western target rises approximately one inch, and the projectile (moving along the tangent) hits it one inch low.

7 Finally, the reverse would happen for the eastern target: at the target, the new horizon $W'E'$ would be one inch below the tangent WE; the target's later position T_E' would be one inch below its earlier position T_E; and the projectile would hit one inch high.

This calculation seems to correspond to my own, formulated in modern notation:

arc TT_W'
= (15 seconds/360 degrees)(length of circumference with radius of 500);
= (15/[(360)(60)(60)])(2π500);
= 15π/[(36)(36)] = 5π/[(12)(36)];
= approximately π/(12)(7) = π/84 = approximately 1/25.

In conclusion, even if one agrees with the anti-Copernicans by making several assumptions that would suggest a vertical deviation for point-blank gunshots, the deviation would be of the order of one inch for a range of about 500 cubits. This amount does seem small enough to entitle one to say that we do not really know whether point-blank gunshots hit on target to that level of accuracy. Thus, by such considerations we cannot disprove that we are on a rotating earth. In other words, the argument from point-blank gunshots is groundless, insofar as it assumes the premise that experience shows that such gunshots hit within one inch of the target, but this premise is groundless.

IIB10. FLIGHT OF BIRDS (DML 212–18)

Besides falling bodies and the motion of projectiles, the flight of birds was another phenomenon which the anti-Copernicans thought created a difficulty for the earth's motion. Galileo states this argument as follows:

> There remains for me only that doubt which I hinted at before, about the flight of birds. Since these have the [animate] faculty of moving at will in a great many ways, and of keeping themselves for a long time in the air, separated from the earth and wandering about with the most irregular turnings, I am not entirely able to see how among such a great mixture of movements they can avoid becoming confused and losing the original common motion. Once having been deprived of it, how could they make up for this or compensate for it by flying, and keep up with all the towers and trees which run with such a precipitous course toward the east? I say "precipitous," because for the great circle of the globe it is little less than a thousand miles an hour, while I believe that the swallow in flight makes no more than fifty.
>
> (DML 212–13)[10]

Before criticizing this argument, Galileo points out that it seems stronger than the previous anti-Copernican arguments because the previous criticisms do not seem to apply to it.

For example, it would be wrong to object that if the earth were rotating then the lower regions of the atmosphere would be rotating with it, and so flying birds would be able to keep up with terrestrial rotation through the agency of the air. This objection is wrong because it does not seem that the moving air could impart to a solid body a degree of speed equal to its own; the agency of the air would be insufficient to accomplish this much (DML 213–14).

Galileo advances a second appreciation of the argument. The key point is that birds have the power of self-movement, which gives them the ability to move at will against the innate tendencies of terrestrial bodies; for example, when dropped from the top of a tower, a dead bird will fall to the ground like any heavy body, whereas a live bird will fly away. Thus, the principles that apply to inanimate bodies like projectiles and falling bodies do not seem to apply to animate ones like birds (DML 216, FAV 212.3–10).

Despite these merits, the birds argument is incorrect. First (DML 214), although on a rotating earth the motion of the air would not, by itself, be sufficient to carry birds along, it would contribute to them retaining the rotational motion which they would possess insofar as they are, in part, terrestrial bodies.

Second (DML 216, FAV 212.10–30), although it is true that animate bodies like birds do not obey all the principles which inanimate bodies like projectiles obey, the two sets of bodies do share some principles. A key difference between the two is that birds have an internal cause of movement, whereas projectiles have an external cause; and a key commonality is that the motion of both obeys the principle of superposition of motion. Thus, on a rotating earth birds' ability to fly and power of self-movement would amount to being able at will to add or subtract degrees of speed to the fixed rotational speed deriving from the earth.

Finally (DML 216–18), there is a crucial experiment that provides a conclusive refutation of the birds argument, and also of all the others from falling bodies and projectiles. The experiment is made on a ship in a cabin under deck where one has the following arrangement: some water is dripping into a bottle on the floor from a container up above; some incense is burned and the smoke is allowed to rise up freely; some fish swims around in a bowl of water; some flies and butterflies have been released and allowed to move around the cabin at will; some people throw a ball to each other in various different directions; and other people jump in all directions with both feet together. The experiment is made at two different times, when the ship is standing still and when it is moving forward at some constant speed. Observation reveals that there is no difference deriving from the state of rest or uniform motion of the ship, and that the flies, fish, and butterflies share the motion of the ship as easily and effortlessly as the inanimate bodies.

IIB11. EXTRUDING POWER OF WHIRLING (DML 218–53)

Another argument against the earth's motion was based on the fact that in a rotating system, or in motion along a curve, bodies have a tendency to move away from the center of rotation or of the curve. The argument added that if the earth rotated, bodies on its surface would be traveling in circles around its axis at different speeds depending on the latitude, the greatest speed being about 1,000 miles per hour at the equator. This very high rate of speed would presumably generate such a strong extruding tendency that all bodies would fly off the earth's surface, and the earth itself might disintegrate. Since this obviously does not happen, the earth must not be rotating.

Using Galileo's own terminology, this objection may be labeled the argument from the extruding power of whirling, or the extrusion argument. Such extruding power corresponds to what modern physics calls centrifugal force or tendency, and so, anachronistically speaking, it could also be labeled the argument from centrifugal force. This argument is not found in Aristotle's works, but there is a cryptic statement of it in Ptolemy's *Almagest*; Copernicus attributes the argument to Ptolemy, but manages to give a slightly clearer formulation; and Galileo apparently follows Copernicus in attributing it to Ptolemy.[11]

However, with Galileo the argument receives the attention it deserves. His initial formulation reads as follows:

> Here is another very ingenious argument, taken from the following observation; it is this: circular motion has the property of extruding, scattering, and throwing away from its center the parts of the moving body whenever the motion is not very slow or the parts are not

attached together very firmly. For example, consider those huge treadmill wheels designed so that the walking of a few men on their inner surface causes them to move very great weights, such as the massive rollers of a calender press or loaded barges dragged overland to move them from one river to another; now, if we made one of these huge wheels turn very rapidly and its parts were not very firmly put together, they would all be scattered along with any rocks or other material substances however strongly tied to its external surface; nothing could resist the impetus which would throw them with great force in various directions away from the wheel, and consequently away from its center. If, then, the earth were rotating with a very much greater speed, what weight and what strength of mortar or cement would keep rocks, buildings, and entire cities from being hurled toward the sky by such a reckless turning? And think of people and animals, which are not attached to the earth at all; how would they resist so much impetus? On the contrary, we see them and other things with much less resistance (pebbles, sand, leaves) rest very calmly on the earth and fall back to it even when their motion is very slow.

(FIN 154; cf. DML 153–54)

This formulation is a great improvement over previous versions. However, Galileo thinks that other improvements are needed before refuting the argument. One is to strengthen its crucial premise, by giving further evidence to establish the reality of the extruding power of whirling. For example, he mentions the experiment of tying a small pail of water at the end of a string and whirling the pail in a vertical circle by the motion of one's hand; then suppose a small hole is made in the bottom of the pail; as the pail is whirled around one will see water rushing out of the hole always in a direction away from one's hand. When Galileo introduces these considerations, he does it with words that leave no doubt about his intention: "I want to strengthen and tighten it further by showing even more sensibly how true it is that, when heavy bodies are rapidly turned around a motionless center, they acquire an impetus to move away from that center, even if they have a propensity to go toward it naturally" (FIN 173; cf. DML 220).

Even more striking is that fact that Galileo begins his discussion with an essential clarification. He points out that as usually formulated (and as formulated in the passage just quoted), the argument is improperly stated. That is, its crucial step should be stated to say that if the earth were rotating then there would now be no loose bodies on its surface, since they would have all been extruded long ago due to the extruding power of whirling. Instead, the argument is usually misstated by claiming that if the earth were in rotation then we would see bodies on its surface extruded off toward the sky. In Galileo's own words:

> This refutation refers to the destruction of buildings and to rocks, animals, and men themselves being cast toward the heavens; but such destruction and scattering cannot happen to buildings and animals that do not already exist on the earth, nor can men be born and buildings erected on the earth unless it stands still; therefore, it is clear that Ptolemy is arguing against those who grant the earth to have been at rest for a long time (that is, while animals and masons could live on it, and palaces and cities could be built), but then suddenly set it in motion, to the ruin and destruction of buildings, animals, etc. On the other hand, if he had intended to argue against those who attribute this turning to the earth since its original creation, he would have refuted them by saying that, if the earth had always been in motion, then neither beasts nor men nor rocks could ever have been formed on it, and still less could buildings have been erected and cities founded, etc. ... [In other words,] Ptolemy argues either against those who regard the earth as being always in motion, or against those who regard it to have been still for some time and then to have been set in motion. If he is arguing against the first, he should have said: "the earth has not always been in motion because terrestrial rotation would not have allowed men, animals, or buildings to exist on the earth, and so there would have never been any of them on it." However, he argues by saying: "the earth does not move because the beasts, men, and buildings already found on the earth would be cast off"; hence, he supposes the earth to have been once in such a state as to have allowed beasts and men to form and live on it; this implies the consequence that once it was motionless, namely, suitable for animal life and the construction of buildings.

> (FIN 172; cf. DML 219–20)

These remarks can also be interpreted as a defense of the extrusion argument from an objection. The objection would be that, as ordinarily stated (and as stated in the first formulation above), the argument alleges to be proving one conclusion (that the earth is not in motion), but instead, at best, proves another (that the earth did not recently begin to move). That is, the argument reaches an irrelevant conclusion, namely, a proposition that is not being disputed; and so it is an instance of the fallacy catalogued by Aristotle (*On Sophistical Refutations*, 167a21) under the label of *ignoratio elenchi*. Galileo's defense would be that such a criticism is self-defeating because, as soon as it is made, it becomes immediately obvious how the original argument should be stated and was intended to be stated.

After all these improvements, clarifications, and appreciations, Galileo proceeds to refute the argument. His refutation of this argument consists of three criticisms.

First (DML 222–29), if a body were to be extruded from a rotating earth, the extrusion would occur along a tangent to the point of last contact with the terrestrial surface; the reason for this stems from the principle of conservation of motion. But, because of gravity, on a rotating earth bodies would still have a tendency to move downward along the secant from the point of their position to the center of the earth. Thus, we must compare these two tendencies; we cannot just consider the centrifugal extrusion, as the anti-Copernican argument seems to be doing. Now, the comparison shows that the downward tendency is greater than the extruding one. For example, consider Figure 5.5 (DML 230), where the circle EHF represents the equator of the earth rotating clockwise; while a body would have the tendency to be extruded along the line HG, it would also have the tendency to fall along the line GED.

The only thing wrong with this criticism is that it is somewhat incomplete and not as detailed as it might have been. Galileo does suggest that, near the point of tangency, the downward tendency would be 1,000 times greater than the extruding one (DML 226–27). The number 1,000 is meant as a rough estimate, as a claim about the order of magnitude. At this point, Galileo could have been more quantitatively precise about the comparison.

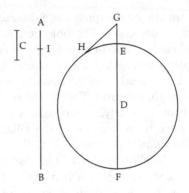

Figure 5.5

For example, he could have calculated the distance covered in one second by a body falling from rest; he had done extensive research on this topic, and in the next section (IIC1) he gives some relevant approximate figures (i.e., 100 cubits in 5 seconds). Then he could have calculated the distance separating the extrusion tangent and the terrestrial circumference in one second of time at the equator; this can be computed from knowing the earth's radius, the rate of terrestrial rotation, and the geometry of the situation. He could have easily arrived at the number 266 as the ratio between these two distances; that is, at the conclusion that in one second free fall takes a body 266 times farther downwards than terrestrial rotation extrudes it away from the center. In fact, a few years after the publication of the *Dialogue*, Marin Mersenne (working directly from Galileo's book) used this method of calculation and arrived at essentially this number (cf. MacLachlan 1977: 176–78). Later in the century, Huygens and Newton arrived at more adequate solutions of the problem, corresponding to modern physics; that is, at the equator the centripetal acceleration due to gravity is 289 times greater than the centrifugal acceleration due to terrestrial rotation, which is to say that a body weighs 1/289 less at the equator than it would on a motionless earth (cf. Chalmers and Nicholas 1983: 322).

However, Galileo did not perform this type of calculation. Though his motivation is unclear, one likely reason is that he thought he could prove something much stronger than the contingent fact that the downward tendency due to weight exceeds the extruding tendency due to rotation; his stronger claim is that the downward tendency (however small, as long as it is not zero) can always overcome the extruding tendency (however large). This is the gist of his next explicit criticism, to which I now turn.

In his second criticism (DML 229–35), Galileo argues that the downward tendency not only happens to exceed the tangential one, but necessarily does so for mathematical reasons; i.e., that extrusion would be mathematically impossible on a rotating earth. He tries to prove this mathematical impossibility based on the geometry of the situation in the neighborhood of the point of contact between a circle and a tangent, and on the behavior of the external segments (called exsecants) of the secants drawn from the center of the circle to the tangent, as shown in Figure 5.6 (DML 231; cf. MacLachlan 1977: 175). He follows three lines of reasoning: as one approaches the point of tangency, (1) the ratio of an exsecant to the corresponding tangent segment tends toward zero; (2) the ratio of an exsecant to the corresponding speed of fall also tends toward zero; and (3) also tending to zero is the ratio of one exsecant to another at twice its distance from the point of tangency. That is, the exsecants get smaller and smaller in relation to (1) the corresponding tangent segments, (2) the speeds of fall, and (3) each other. Thus, the distances of fall required to prevent extrusion become infinitely smaller than the distances required to achieve extrusion, and they decrease more rapidly than the speeds of fall.

Galileo's three-pronged proof contains parts that are mathematically valid but are misapplied to the physical situation, and parts that would be physically applicable but are mathematically incorrect. However, such details are beyond the scope of this book.[12]

On the other hand, another issue is worth exploring here. One of the most common reactions to this Galilean argument is that there must be something wrong with it because the conclusion is false, i.e., the physical claim that on a rotating earth bodies could not be extruded; for in fact, if the earth rotated faster or gravity were weaker, bodies would be extruded. Even if this criticism of Galileo were correct, it would only be the beginning of a critical analysis of his argument because all it would tell us is that either

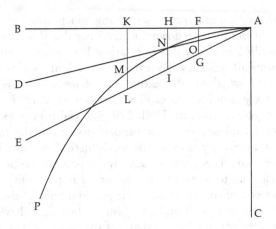

Figure 5.6

some of his initial premises are false or some of his principles of inference are invalid; then we would have to reconstruct his argument in all its complexity to determine this. However, let us examine whether this common criticism is really correct. Is it the case that on a rotating earth bodies with weight could be extruded? I believe that, from the point of view of modern physics, Galileo's claim is indeed false in one sense, but it is true in another.

Two meanings of extrusion must be distinguished, and it can be shown that bodies could be extruded in one sense, but not in the other. The two types of extrusion are: (a) leaving the earth's surface and going into a geocentric orbit; and (b) escaping from the earth's gravitational field.

Regarding orbital extrusion (a), to see that it could happen as a result of (increased) terrestrial rotation, we may reason as follows: (1) when a body moves in a circle, a centripetal force is required, otherwise it would move in rectilinear uniform motion (because of inertia); (2) the centripetal force is directly proportional to the square of the linear speed and inversely proportional to the radius ($F = mV^2/R$); (3) for terrestrial bodies this force is provided by the earth's gravitational attraction, which is inversely proportional to the square of the radius, but does not depend on speed ($F = GmM/R^2$); thus, (4) as the rate of rotation increases, the required centripetal force also increases, but the available gravitational attraction remains constant; so (5) the point would be reached when the former exceeds the latter and orbital extrusion occurs; (6) the minimum orbital speed V_o is such that $(mV_o^2/R) = (GmM/R^2)$, or $V_o^2 = GM/R$; but (7) once the body is in orbit, any increase in terrestrial rotation does not affect its velocity or the required centripetal force, and so the body simply remains in orbit and does not escape.

Regarding escape extrusion (b), we may reason as follows: (8) to escape, a body must be given enough speed so that it will never be pulled back to earth, namely so that it can reach an infinite distance from the earth; that is, (9) its kinetic energy must be sufficient to do the work required to move it from the earth's surface to infinity; (10) this work is equal to the body's change in potential energy; (11) setting potential energy to zero at infinity, this change equals the body's initial potential energy; that is, (12) the escape velocity is such that $[(mV_e^2)/2] = GmM/R$; or, (13) $V_e^2 = 2V_o^2$; that is, (14) the escape velocity is greater than the orbital velocity (by a factor of the square root of two); so (15) the escape velocity will never be reached by just increasing terrestrial rotation.[13]

Thus, I am not sure Galileo's conclusion is physically wrong; it depends on which of these two physical claims we attribute to him. Therefore, the truth-value of the conclusion does not offer us a clear clue about the argument's correctness.

Galileo's third criticism may be reconstructed as follows (DML 245–53). It is true that the extruding tendency increases as the speed when the radius is constant; but when the speeds are equal, the extruding tendency decreases as the radius increases; so the extruding tendency increases directly with the speed but inversely with the radius; thus, the extruding tendency remains constant when the linear speed increases as much as the radius, i.e., when equal numbers of rotations are made in equal times. Hence, terrestrial rotation would cause as much extrusion as a wheel rotating just once in twenty-four hours; so, a rotating earth would not scatter bodies toward the heavens. To show that at constant speed the extruding tendency decreases as the radius increases, one may reason thus: the extruding tendency is equal to the force needed to prevent the body from escaping along the tangent; this force is equal to that required to compensate for the tangential displacement if the body were extruded; and this compensation is smaller for greater circles. The reason for this last claim can be seen from Figure 5.7 (DML 250, FIN 209): when the linear speeds are the same (arc CE = arc BIG), the deviations from the tangent are measured by the exsecants (DE and FG), and the exsecants become smaller as the circle becomes larger (DE < FG).

This Galilean criticism is partly right but partly wrong. The linear speed of a body lying on the earth's equator, which is about 24,000 miles long, is about 1,000 miles per hour. The original anti-Copernican argument seems to assume that the extruding tendency varies as the linear speed, so that the speed of 1,000 miles per hour should generate an extruding tendency that would tear the earth apart. Galileo's key critical claim here is that this assumption is not right, that it is incomplete; it fails to take into account another crucial variable—the radius of rotation; this is a serious flaw because the radius affects the extrusion in an inverse manner; that is, the extruding tendency decreases as the radius increases. From the viewpoint of modern physics, this Galilean claim is essentially right.

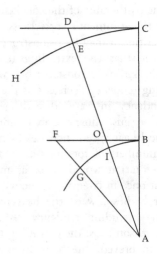

Figure 5.7

However, the manner in which Galileo derives this claim is problematic in several ways, which need not be discussed here.[14] Moreover, the situation is even more complicated than he realized: his critical claim is itself incomplete because he failed to realize that although the extruding tendency does grow with the linear speed, the growth is in accordance with the square of this speed; in short, centrifugal force is directly proportional to the square of the linear speed and inversely proportional to the radius; thus, the extrusion is a function not simply of the angular speed, but of the radius and the *square* of the angular speed.

Even so, Galileo's discussion embodies his intuition that the extruding tendency depends more on angular speed than on the radius, which is one way of stating one consequence of the correct law of centrifugal force ($F = m\omega^2 R$).

In sum, Galileo's discussion of the extrusion objection contains both a constructive and a destructive part. The constructive discussion contains an improved statement, strengthening, clarification, appreciation, and defense of the argument; these are intended to show that this is a serious difficulty that cannot be dismissed but must be overcome. The destructive discussion attempts to show that this objection is quantitatively invalid: the extruding power of bodies on a rotating earth would be less than their downward tendency, as a matter of contingent fact; it would necessarily be less, as a matter of mathematical necessity; and it would be quite small because the angular speed would be very small. Finally, the destructive criticism has a constructive aspect from the point of view of the nature of circular motion, for here Galileo is groping his way toward the physical laws of circular motion.

IIC1. FALLING TO EARTH FROM THE MOON (DML 253–71)

The anti-Copernican arguments examined so far can be traced directly or indirectly to Aristotle, Ptolemy, and Tycho. Next, Galileo examines some arguments found in more recent writings by authors who were his contemporaries. The first of these works is a book entitled *Mathematical Disquisitions on Controversial and Novel Astronomical Topics* (1614), written by Johann Georg Locher, under the supervision of his mentor, the German Jesuit, Christoph Scheiner. Galileo begins by focusing on an argument which may be regarded as a novel version of the argument from vertical fall. Simplicio summarizes Locher's argument as follows:

> He takes many pains in calculating exactly how many miles an hour are travelled by a point of the terrestrial globe situated under the [celestial equator] and how many miles are passed by other points situated in other parallels; and ... how many miles a cannon ball would go in the same time, being placed in the concave of the lunar sphere, supposing it also as big as Copernicus himself represents it, so as to take away all subterfuges from his adversary ... he shews that a heavy body falling from thence above would consume more than six days in attaining to the centre of the Earth ... it is in his, and also in my opinion, a most incredible thing that, in descending downwards, it should all the way maintain itself in our vertical line, continuing to turn round with the Earth about its centre for so many days, describing under the [equator] a spiral line in the plane of the great circle itself; and under other parallels, spiral lines about cones; and under the poles falling by a simple straight line. In the next place, he establishes and confirms this great improbability by proving ... many difficulties impossible to be removed by the followers of Copernicus ... supposing that the velocity of the body falling along the vertical line towards the centre of the Earth were equal to the velocity of its circular motion in the grand circle of the concave of the lunar sphere ... it would pass in an hour 12,600 German miles, a thing which indeed savours of impossibility. Yet, to shew his abundant caution, and to give all advantages to his adversaries, he supposes it for true and concludes that the time of the fall ought still to be more than six days.
>
> (Santillana 234–35; cf. DML 254–55)[15]

This argument seems to have three main parts. A key part is the computation that it would take six days for a terrestrial body falling to earth from a height equal to the earth–moon distance, on the assumption that its speed of vertical fall would be equal to the eastward linear speed of diurnal rotation at that height. Based on this computation, another part of the argument claims that if the earth rotated and bodies falling freely from any height appeared to move vertically (i.e., were seen by a terrestrial observer to be moving perpendicular to the earth's surface from the point of release), then bodies falling from the moon would have to turn around the earth's center six times before arriving there six days later, as well as traverse complicated spiral trajectories. Third, Locher claims that it is impossible that bodies falling from the moon could keep up with the earth's rotation for such a long time, for various reasons the gist of which is that there is nothing that could account for them moving in this strange manner.

Galileo begins his criticism by arguing that Locher's computation is mathematically incoherent. The reason is this. The body's original linear speed before being released (allegedly 12,600 mph) is the speed that enables it to move, in the twenty-four hours of one diurnal rotation, over the circumference of a circle whose radius is the distance between the earth and the moon; now, to fall back to earth, the body has to move over a distance equal to the radius of this circle; but given Locher's own assumption that the speed of vertical fall is the same as the linear speed along the circumference before release, then the body will fall to earth in about 1/6 of one day (or about four hours), since geometry tells us that the radius of a circle is about 1/6 of the circumference (i.e., circumference = 2π times the radius). The incoherence is that Locher is saying that a body moving at the same constant speed will cover the circumference of a circle in one day, and its radius in six.

Galileo finds it hard to believe that anyone could have committed such a blunder, and so he has the speakers in the dialogue double-check Locher's text. The consultation reveals that indeed Locher is guilty of such a mathematical self-contradiction. His miscalculation overestimates the time of fall from the moon by a factor of 36 (given his own assumptions). In attempting to explain the source of Locher's error, Galileo has the speakers mention the possibility that perhaps mathematical propositions are true only in the abstract, but do not correspond to the concrete world in reality. But even Simplicio rejects this explanation in this case. My own explanation[16] is to say that Locher inadvertently reversed the numbers, in the sense that he should have said that the time of fall would be six times shorter than the time required to cover the circumference, but ended up saying that the former would be six times longer than the latter.

However, Galileo does not linger on this criticism. Instead, he stresses a factual criticism, since this gives him the opportunity to discuss, however briefly, the laws of falling bodies which he had discovered, and the new science of motion on which he had been working for a long time.

That is, Galileo objects to Locher's arbitrary assumption about the value of the body's speed of fall. For it disregards the fact that the speed of falling bodies is accelerated, not constant; it also disregards the value of this acceleration, i.e., the proportion in which the speed of fall increases. There are different equivalent ways of expressing this relationship. One is to say that the acceleration is constant, or that the speed increases uniformly, or that the instantaneous speed is proportional to the time elapsed. Another way is the law of odd numbers: in free fall, the distances covered in successive equal times increase as the odd numbers from unity (1, 3, 5, 7, etc.). Third, there is the law of squares: the distance fallen from rest increases as the square of the time elapsed. Implicit in these laws, but explicitly stated by Galileo, is also the idea that the speed of fall is independent of weight (if we disregard external disturbances); that is, all bodies fall at the same rate, regardless of weight. Galileo is keen on stating this principle explicitly in order to expose the error of Aristotle, who held that heavier bodies fall faster, in proportion to their weight. In this context, Galileo is also keen to briefly discuss another idea, involving the (approximate) isochronism of the pendulum: that is, the frequency of oscillation of a pendulum is independent of the amplitude of the oscillation, and depends only on the length of the pendulum. Finally, Galileo also finds the pretext to elaborate (and even attempts to demonstrate) the so-called double-distance rule: that the speed acquired by a falling body during a certain period of uniformly accelerated fall from rest is such that, if the body were to move uniformly at that final speed, then during an equal period of time it would cover a distance double that already covered.

The law of squares is exploited by Galileo to compute the time of fall from the moon. But to do this, he also needs some empirical data about the actual time taken by a body to fall a particular distance. He states having found that "from repeated experiments an iron ball of one hundred pounds falls from the height of one hundred cubits in five seconds" (FAV 250; cf. DML 259). Then he goes through the computation in detail, and derives that the time of fall from the moon would be: 3 hours, 22 minutes, and 4 seconds. This result shows the factual incorrectness of Locher's computation of six days.

Furthermore, the double-distance rule is exploited by Galileo to advance what may be called a rhetorical criticism of Locher's argument. Such criticism is legitimate because Locher's argument claimed to have the merit of giving the Copernican opponents various advantages, such as using their estimate of the earth–moon distance and assuming for the falling body an absurdly large speed of fall; thus, here Galileo is countering Locher's rhetorical strategy. In fact, the double-distance rule tells us that the final speed acquired by the body falling to earth from the moon would be greater than the speed Locher is assuming.

To see this, let us round off the time of fall computed by Galileo to 3 1/2 hours. Applying the double-distance rule to this case, we get that when after 3 1/2 hours of accelerated fall the body reaches the center of the earth, it has a speed which if kept constant would enable the body to traverse in 3 1/2 hours double the distance it has fallen; now, it has fallen a distance equal to the radius of the circle over which it was moving with diurnal rotation before being released; thus, the acquired speed would enable the body to traverse the diameter of that circle in 3 1/2 hours; so the same speed would enable it to traverse the whole circumference of that circle in about 11 hours. Such a speed is about twice as fast as the speed assumed by Locher, since the latter, being the speed of diurnal rotation at the height of the moon, merely enables the body to traverse that circumference in 24 hours.

This concludes Galileo's criticism of the first part of Locher's argument, which is thus shown to be flawed mathematically, empirically, and rhetorically. However, it should be noted that the Galilean criticism has itself a few blemishes, at least from the point of view of modern physics.

First, Galileo is presupposing that the acceleration of falling bodies is universally constant, even over large distances such as falling to earth from the moon. However, we know from Newton's law of gravitation that the acceleration of falling bodies varies (inversely) as a function of (the square of) the distance, since the acceleration derives from the gravitational force, which varies in the same manner. Now for small changes in distance, such as a body falling to the ground from the top of a tower, the change is small enough that it can be regarded as non-existent for most practical purposes. However, for the distances involved in Locher's experiment, the change would be so large that it could not be disregarded.

Moreover, the empirical value of the acceleration used by Galileo is about half the true value. For if we take 1 cubit to equal about 56 cm, then his value of 100 cubits in 5 seconds translates to 5,600 cm in 5 seconds; by the law of squares, this becomes 224 cm in 1 second; and this represents an acceleration of 448 cm/sec^2. However, modern physics tells us that the acceleration near the earth's surface is 980cm/sec^2, which yields a distance of 490 cm in the first second (cf. Beltrán-IT 498–99).

These comparisons are useful to keep in mind, but should not be over-emphasized. For I do not think they undermine the power of Galileo's criticism. Moreover, to stress them unduly would prevent us from appreciating the fact that the more correct facts, figures, and principles of modern physics became possible basically by pursuing further Galileo's own facts, figures, and principles, as well as his approach to the search for truth.

This qualification applies also to another criticism which Galileo advances against Locher's argument, specifically directed to what I have called the third part of his argument. This was the part claiming that bodies falling to a rotating earth from the moon would not be able to keep up with its rotation and would lag behind, especially during such a long period as six days. And now Simplicio adds that the same doubt would remain even if one uses the time computed by Salviati, namely 3 1/2 hours, which is presumably long enough to make the falling body lag behind by a significant amount. Galileo's reply to this point involves basically an application of the principle of conservation of motion. But in this case he was perceptive enough to see that the body falling from the moon would conserve its eastward linear speed of diurnal motion, not its angular speed, with the result that the body would not only not lag behind, nor even descend vertically, but rather advance forward and deviate to the east. In short, on a rotating earth, bodies would not only not fall vertically in reality, but also not even appear to fall vertically, at least if the distance fallen was great enough to make the forward advance (or eastward deviation) observable or measurable. This intuition represents an anticipation of subsequent developments in modern physics, such as the coriolis effect, which I have similarly mentioned before, in the discussion of north-south gunshots (IIB8). Thus, the relevant passage deserves to be quoted:

> Your author supposes that the ball, while it was on the surface of the sphere of the Moon, would participate in the circular motion every twenty-four hours which it shares with the Earth and everything else within the lunar sphere; in which case the same force that impelled its circular motion before it started to fall would continue to affect it while it was falling. Then, far from failing to keep up with the Earth's motion and being left behind, it ought rather to run ahead of it, because as it approached the Earth it would be revolving in continually decreasing circles; so if the ball maintained the same velocity that it had when it was on the surface of the sphere, it would run ahead of the rotation of the Earth.

> (Shea-Davie 297; cf. DML 271)

The third part of Locher's argument also advances a more general reason why bodies falling to a rotating earth from the moon could not follow the complicated trajectories elaborated by him, more general than his claiming that such bodies would lag behind during such a long descent; the more general reason is that there is nothing that could explain or account for their following such trajectories. So far Galileo has not criticized this more general argument. But that is the main subject of the next passage.

IIC2. INEXPLICABILITY OF TERRESTRIAL ROTATION (DML 271–86)

As mentioned, part of Locher's argument claims that bodies falling to a rotating earth could not rotate with it because there would be no way to explain their rotation. From the perspective of the concrete problems discussed in the previous section, this inexplicability is a relatively minor issue. However, from the perspective of epistemology, this issue is important. For the same would, allegedly, apply directly to terrestrial rotation itself; that is, if the earth rotated, there would be no explanation of why this rotation was taking place; and since Locher is concluding that the earth does not rotate, he is assuming that inexplicability implies non-existence: that if some hypothetical situation is such that its occurrence could not be explained, then it cannot exist. Thus, we may call this anti-Copernican objection the argument from the inexplicability of terrestrial rotation. Referring to Locher, Simplicio summarizes the argument in the following passage:

> The Author raises an objection, as you see, demanding on what principle this circular motion of heavy and light bodies depends, that is whether upon an internal or an external principle ... The Author proves that it can be neither inward nor outward ... but if heavy and light bodies can have no principle of moving circularly, either internal or external, then neither can the terrestrial globe move with a circular motion, and thus you have served the intent of the Author.
>
> (Santillana 250; cf. DML 271)[17]

Galileo's criticism is three-fold. First (DML 271–72), even if it were the case that we could not explain the earth's rotation, this would only mean that we do not know the explanation of the earth's rotation; and from this it does not follow that the earth's rotation is false. That is, here Galileo is rejecting the epistemological principle presupposed by this argument, that inexplicability implies non-existence; and he is giving a reason for denying the principle, namely that inexplicability implies ignorance of the cause, but such ignorance obviously does not imply non-existence.

Second, if and to the extent that the earth's rotation is inexplicable, it is no more inexplicable than the fall of heavy bodies, the motion of projectiles, or the diurnal motion of heavenly bodies in the Ptolemaic system. In fact, it is no explanation for the Aristotelians to call the causes of these phenomena, respectively, gravity, impressed virtue, and assisting intelligence; for these are mere names, and knowledge of the name is not the same as knowledge of the cause or essence, since the latter requires understanding, but naming does not. Here, Galileo is suggesting that explicability and understanding are relative notions such that judgments involving them require a critical comparison among things that are understood, things that are not understood, and things that are more or less understood. His words are memorable enough to deserve quotation:

> *Salviati* ... If this author knows by which principle other world bodies are moved circularly, ... then I say that that which makes the earth move is something similar to whatever moves Mars and Jupiter, and which he believes also moves the stellar sphere. If he will advise me as to the mover of one of these movable bodies, I promise I shall be able to tell him what makes the earth move. Moreover, I shall do the same if he can teach me what it is that moves earthly things downward.
>
> *Simplicio.* The cause of this effect is well known; everybody knows it is gravity.
>
> *Salviati.* You are wrong, Simplicio; what you ought to say is that everybody knows it is called "gravity." But what I am asking you for is not the name of the thing, but its essence, of which essence you know not a bit more than you know about the essence of whatever moves the stars around, except for the name that has been attached to it and that has been made a familiar household word by the continual experience which we have of it daily. We do not really understand what principle or what force it is that moves stones downward, any more than we understand what moves them upward after they leave the thrower's hand, or what moves the moon around. We have merely, as I said, assigned to the first the more specific and definite name "gravity"; to the second the more general term "impressed virtue"; and to the last the word "intelligence" (either "assisting" or "informing"); and as the cause of infinite other motions we give "Nature."
>
> (FAV 260–61; cf. DML 272)[18]

Third, Galileo objects that Locher's claim that the earth's rotation would be inexplicable is groundless, in the sense that his subargument intended to establish this claim leaves much to be desired. The subargument amounts to an attempt to show that the explanatory principle would be neither internal nor external; and it could not be internal because it could be neither an internal accident nor an internal substance; and regarding the possibility of its being an external principle, both sides agree in rejecting that possibility.

The details of this discussion need not concern us, but the following points are worth noting. Galileo finds the distinction between internal and external causes problematic because there are cases where the cause of motion would be both. For example, normally the cause of bodies falling downward is taken to be internal, namely some innate tendency of terrestrial bodies; whereas the cause of a projectile moving upward after being thrown is regarded as external, in the sense that its impulse derives from the thrower. Now, imagine a well so deep that it reaches not only the earth's center, but also the point on the earth's surface diametrically opposite; if a rock is dropped into the well, it seems clear that it would move downward with accelerated motion toward the center of the earth; but once this point is reached, it seems equally clear that the body would not stop, but would continue to move; but such continued motion would be upward and retarded, since it would be taking the body away from the earth's center. Let us now ask whether the cause of such continued motion would be internal or external. It could be regarded as internal since it seems clear that the body would spontaneously do that; but it could be regarded as external because such motion would depend on the previous downward motion. In short, in motion along a tunneled earth, the common distinction between internal and external cause does not apply. Additionally, a body moving with such a motion would start to oscillate back and forth along the length of the tunnel on both sides of the earth's center, and such an oscillation would be analogous to the motion of a pendulum; thus, even the common pendulum is a case where the distinction between internal and external cause is difficult to apply.

Moreover, Locher thinks that the cause of the earth's rotation could not be internal, partly because that would imply that the corresponding circular motion would belong intrinsically to bodies of many different natures, e.g., heavy and light bodies, falling bodies and projectiles, and animate and inanimate bodies. Galileo replies that this consequence would indeed follow, but that he sees no difficulty with it. Rather, he finds Locher's reasoning here as absurd as believing "that if a dead cat falls out of a window, a live one cannot possibly fall too, since it is not a proper thing for a corpse to share in qualities that are suitable to the living" (DML 277).

Locher also argues that the principle of terrestrial rotation could not be internal because it would imply that bodies of the same nature would be moving in very different ways. For example, as previously mentioned, if a rock were falling over the equator its motion would be along a plane spiral; if over the poles, its falling motion would be straight; and at other latitudes, the falling motion would be along conical spirals. Galileo replies that again these consequences would indeed follow, but he fails to see any difficulty with them; all that would be happening is that bodies of the same nature would be moving in different ways under different conditions. Locher's error this time seems to be to regard necessary consequences as damaging absurdities (DML 282).

In sum, the argument from the inexplicability of terrestrial rotation is invalid, insofar as even if terrestrial rotation were inexplicable, that would not make it false. The argument is also groundless, insofar as its key premise is questionable: first, terrestrial rotation is no more inexplicable than falling motion, projectile motion, or the motion of heavenly bodies, which is also to say that it is as explicable as they are; and second, it is erroneous to argue, as Locher does, that terrestrial rotation is inexplicable because its cause could be neither internal nor external.

IIC3. DECEPTION OF THE SENSES (DML 287–98)

After examining some of the arguments in Locher's *Disquisitions,* Galileo goes on to discuss some of those in Scipione Chiaramonti's *On the Three New Stars that Appeared in 1572, 1600, and 1604* (1628). The first of Chiaramonti's arguments to be considered is of the same type as the one by Locher criticized in the last section, in the sense that they both primarily raise issues of an epistemological nature. The present argument is based on the important role played by the senses in the acquisition of knowledge. The gist of the argument is that since we have no sensory experience of the earth's motion, if the earth were in motion, that would mean that the human senses were being deceived, which is absurd.

Hence, this objection to Copernicanism may be called the argument from the deception of the senses. It was the most fundamental objection against the earth's motion. In Simplicio's words, which quote and paraphrase Chiaramonti's own, the argument is stated as follows:

"if Copernicus's opinion is accepted, the criterion of natural philosophy seems to be severely undermined, if not completely destroyed." According to the view of all philosophical schools, this criterion requires that the senses and experience be our guides in philosophizing. However, from the Copernican perspective our senses are greatly deceived when our eyes perceive very heavy bodies fall perpendicularly in a straight line without deviating a hairbreadth from it, and they do so in close proximity and in a very clear medium; for, according to Copernicus, that motion is not really straight, but a mixture of straight and circular, and so our vision is deceived in such a clear-cut situation ... The author continues to show how in the Copernican doctrine one must deny the senses and the most basic sensations; for example, although we can feel the breeze of the lightest wind, we do not feel the impetus of a perpetual wind striking us at a speed of more than 2,529 miles per hour; this is the distance which would be covered in one hour by the earth's center in its annual motion along the circumference of the ecliptic, as he diligently calculates[19] ... He continues with the same objection, showing that according to Copernicus one must deny one's own sensations. For, this principle by which we go around with the earth either is intrinsically ours or is external to us (namely a case of being forcibly carried by the earth); if it is the latter, (since we do not feel being forcibly carried) we will have to say that our sense of touch does not feel the very object being touched and does not receive its impression in the sensorium; but if the principle is intrinsic, then we will not be feeling a motion deriving from within us, and we will not be perceiving a propensity perpetually inherent in us.

(FIN 212–17; cf. DML 288–95)[20]

Here, three examples of alleged deceptions of the senses are being considered: seeing freely falling bodies move vertically; no sensation of a strong constant westward wind; and the kinesthetic feeling of rest.

Galileo's criticism consists of a denial that any of these three cases is a genuine instance of deception of the senses, and then of a more general objection to the validity of the argument.

First (DML 288–90), it would be no deception of the senses if on a rotating earth we perceived bodies falling vertically and failed to perceive any lateral component of their motion due to terrestrial rotation. For, as discussed on previous occasions, motion exists only relative to things that do not share it, and so motion shared by an observer and an observed object does not exist for him and is imperceptible to him. But on a rotating earth, the eastward component of the motion of the falling body would be shared by the observer, and hence it would not be there to be perceived. Here Galileo is exploiting the principle of the relativity of motion to elaborate his criticism.

Second (DML 294–95), Galileo objects that there would be no wind deception on a moving earth because there would be no wind for us to perceive; and there would be no wind because wind is, by definition, air moving relative to the observer, and on a moving earth the air as well as the observer would be carried along. At this point Simplicio could question whether the air would indeed be carried along, but doing so would raise issues related to natural motion and the conservation of acquired motion, which were largely discussed earlier; the stress now is on the problem of the deception of the senses, and from this viewpoint Galileo's answer seems cogent.

However, Simplicio, following Chiaramonti, is referring literally to the earth's annual orbital revolution around the sun, rather than to its diurnal axial rotation. And it is puzzling that Galileo should have retained this reference here, given that the "Second Day" is supposed to focus on diurnal motion, and so the talk of annual motion is out of place. On the other hand, thinking distinctly about the two motions, he might have said the following. If the air could not follow the earth's motion, then on the assumption that the earth has the diurnal motion, a perpetual westward wind would follow. But on the assumption that the earth has the annual motion, what would follow is that the air should have been left behind long ago and that there should not be any air on the earth's surface now; so the wind

objection should be reformulated to read "the earth cannot have
the annual motion because if it did then it would now have no
atmosphere; since it is obvious that the earth has an atmosphere,
the earth cannot move annually around the sun"; that is, the
"wind objection" should be restated as the "air objection." This
point is in the spirit of Galileo, being similar to the first critical
clarification he makes to the extrusion objection (IIB11). But
since he does not offer such a critique for the wind objection, he
could be criticized for having failed to do so. In short, either he
should not have injected the annual motion into this discussion at
all (not even by way of quoting from Chiaramonti's book), or
having done so, he should have made the just mentioned critical
clarification.

Third (DML 296–97), our inability to feel the earth's motion
is not a deception either. For our experience with navigation
shows that we can feel only changes of motion and not uniform
motion, and so the earth's constant rotation is not something
susceptible of being felt; thus, the kinesthetic sense that enables
us to feel motion is not being deceived because it is not failing to
detect something that it should be able to detect.

Finally (DML 296–98), Galileo develops a more general criticism that undermines the validity of Chiaramonti's argument;
this flaw would be present even if these three cases considered
were indeed deceptions, or other cases were found. This criticism
is directed at the methodological principle underlying the
objection from the deception of the senses. That is, if there were
sensory deception on a moving earth, that would be no reason
to conclude that knowledge is impossible; the more correct
conclusion would be that knowledge is difficult, that it cannot
rely solely on the senses, and that reason plays an equally crucial
role. And here Galileo elaborates a methodological principle
that may be called critical empiricism and contrasted with
the naïve empiricism of Chiaramonti and the Aristotelians in
general. They make the acquisition of knowledge so dependent
on sensory experience that if the senses are not completely
reliable, then there is no reliable guide in the search for truth
and knowledge is impossible; whereas Galileo stresses that if
the senses are not always reliable, then we should learn to

distinguish situations in which they are reliable from situations in which they are not, and this task can only be performed by reason.

And this leads to another way of expressing this general Galilean criticism. The previous specific critiques showed that the alleged deceptions of the senses are not deceptions based on considerations dealing with the nature of the phenomena in question and the nature of sensation, the key point being that there is no deception in not perceiving something which should not be perceivable. But one can also give a different reason why the alleged deceptions are not deceptions of the senses: they are deceptions of reason (the reason of those who from the fact that certain things appear in a certain way conclude incorrectly that they are really that way).

IIC4. MULTIPLE NATURAL MOTIONS (DML 298–306)

The next argument from Chiaramonti's *Three New Stars* which Galileo examines is one that tries to disprove the earth's motion based on the principle that simple bodies can have only one natural motion (cf. Rosen 1992: 354, note to p. 17 line 22). Given this principle, the Copernican system would be in trouble immediately on account of the two crucial motions which it attributes to the earth: daily axial rotation and annual heliocentric revolution.

However, the situation was even more problematic because Copernicus himself (*On the Revolutions*, bk. 1, ch. 11) had attributed to the earth a third motion, which he (erroneously) thought was needed in order to explain the fact that the earth's axis remains always parallel to itself. That is, for Copernicus, while the earth orbits the sun from west to east in one year and rotates daily on its axis in the same direction, the terrestrial axis (which is inclined to the plane of that orbit) describes the surface of a cone by a motion of precession in the opposite direction. It's as if the earth's inclined axis were rigidly connected to the sun, so that left to itself it would intersect the sun-earth line always at the same angle but constantly change its relation to the celestial sphere; whereas in order to keep the latter relation constant, the axis has to precess in the opposite direction to compensate for the orbital revolution. Of this more below.

Finally, from the Aristotelian perspective, there was the natural motion of the earth, straight toward the center of the universe. This was regarded as unquestionable, since it was theoretically grounded in the Aristotelian cosmos, and allegedly visible in the free fall of heavy bodies.

Although the anti-Copernican consequences of the principle of the impossibility of multiple natural motions are relatively obvious, one might wonder why such a principle should be accepted. The Aristotelians justified it by an argument based on three metaphysical axioms: every event has a cause; nothing can be the cause of itself; and the same causes have the same effects, i.e., different effects have different causes. When such a justification is taken into account, it becomes evident that the argument from the impossibility of multiple natural motion is trying to show that the earth's motion contradicts some of the basic metaphysical principles about cause and effect; in short, it is metaphysically impossible.

All these ideas are elegantly tied together in the following passage, which Simplicio reconstructs from Chiaramonti's book by way of paraphrase and translation:

> The earth cannot move by its own nature in three widely different movements without actually contradicting many manifest axioms.
>
> The first of these is that every effect depends upon some cause; the second, that nothing is self-created; from these it follows that the thing causing motion and the thing moved cannot be one and the same. This holds not only for things which are moved by an extrinsic and obvious mover, but the above principles imply also that the same holds for natural motions depending upon an intrinsic principle. Otherwise, since the moving thing, as such, is a cause, and the thing moved, as such, is an effect, then the cause and the effect would be identical in all respects. Therefore a body does not move entirely of itself.so that the whole is mover as well as moved, but there is required in the thing moved some way of distinguishing the efficient principle of motion from that which is moved with such motion. The third axiom is that in things subject to sensation, one thing, insofar as it is one, produces but one effect. In an animal, to be sure, the soul does produce various operations, but it does so by means of

different instruments, such as sight, hearing, smell, generation, etc.; in a word, it may be seen that different actions in sensible objects derive from differences which exist in the causes.

Now if these axioms are combined, it will be quite evident that a simple body, such as the earth, will by its nature be unable to move with three widely differing motions at the same time. For by the assumptions made, the whole cannot move by itself. Hence three principles must be distinguished for three motions in it; otherwise the same principle would be producing more than one motion. But if a body contained within itself three principles of natural motion, besides the part moved, it would not be a simple body, but one composed of three moving principles plus the part moved. If therefore the earth is a simple body, it does not move with three motions. Indeed, it will not move with any of the motions which Copernicus attributes to it; for it is obliged to move with only one motion, that is, obviously (for the reasons given by Aristotle), toward its center—as shown by particles of earth, which descend to the spherical surface of the earth at right angles.

(FAV 281–82; cf. DML 298–99)[21]

Galileo begins his criticism by questioning the principle that different effects must have different causes. This objection exploits an Aristotelian idea endorsed by Chiaramonti himself, namely that in an animal the soul produces different operations by utilizing different instruments, such as the various senses and faculties of the animal; for here we seem to have a case where the same cause has managed to produce different effects. Something similar could be happening to the case of the earth, so as to have a single principle cause different motions. Galileo strengthens this criticism by elaborating two issues.

First, it is irrelevant for Chiaramonti to defend his argument by saying that animals have joints that allow them to move in different ways, whereas the earth has no such joints, and so the analogy between the two cases does not hold. For the function of joints in animals is not to allow them to move their whole body simultaneously in different ways, but rather to allow them to move some parts while other parts are not moving; indeed there is no conceivable way in which the earth could have joints that would enable it to generate the multiple motions attributed to it by Copernicus, and hence even if it did have joints it still would not have those multiple natural motions.

Second, it is a misconception for Chiaramonti to defend his argument by claiming that the earth's three Copernican motions cannot derive from a simple cause because they are too disparate, indeed the contrary is true. In fact, Chiaramonti asserts that the earth's annual motion would be from west to east; that its diurnal motion would be from east to west; and that the motion of the axis hypothesized by Copernicus would be from north to south for half a year and from south to north for the other half.

Regarding the last motion, Chiaramonti shares Copernicus's own error about the cause why the earth's axis always points in the same direction relative to the celestial sphere—why it always remains parallel to itself. What Chiaramonti adds is a possible way of redescribing the annual precession of the terrestrial axis in a direction opposite to the earth's annual motion; that is, Chiaramonti thinks of this precession as a wobbling from north to south for half a year and from south to north for the other half; and this, for him, compounds the contrarieties and the impossibility of the Copernican motions. However, Copernicus's third motion becomes unnecessary once the rigid connection between the sun and the earth is done away with (together with all the interlocking mechanisms of the crystalline heavenly orbs), and the various heavenly bodies are regarded as loose and freely floating in a fluid; for the constant parallelism of the earth's axis then becomes an instance of rest, a consequence of the fact that the axis does *not* precess.[22] Galileo and other followers of Copernicus understood this very well, but Chiaramonti apparently did not.

However, Chiaramonti betrays a more serious misunderstanding, regarding the directions of the annual and diurnal motions. In the Copernican system these are both from west to east. It is the Ptolemaic system that makes them contrary: diurnal motion from east to west, and annual motion from west to east. What Chiaramonti does not seem to understand is the following, as we saw earlier (Chapter 3.1): in explaining the apparent diurnal and annual motions geokinetically instead of geostatically, the westward direction of the universal diurnal motion has to be reversed into an eastward terrestrial rotation, whereas the eastward direction of the sun's annual geocentric revolution does not have to be reversed into the earth's annual heliocentric revolution.

This first Galilean criticism undermines the principle of the impossibility of multiple natural motions by questioning one of the principles used by Chiaramonti to justify it, namely that different effects must have different causes. Additionally, Galileo points out that, besides being incorrectly justified, the impossibility of multiple natural motions can be refuted by the empirical evidence of the recent astronomical discoveries. Here Galileo refers to the axial rotation of the planet Saturn, the motion of sunspots, and the satellites of Jupiter; these are phenomena revealed by telescopic observation and showing that it is possible for a body to move with more than a single natural motion.

This is most easily illustrated by Jupiter's satellites. With the telescope, Galileo had detected four bodies near Jupiter, and demonstrated that they revolve around it in orbits of different sizes and with different periods. The key point in this context is that these bodies have at least two natural motions. One is their just-mentioned revolution around Jupiter, previously unknown and recently demonstrated by Galileo beyond any reasonable doubt. The second is the motion they all share with Jupiter: for both sides of the controversy agreed that Jupiter is a planet, and that it revolves in an orbit that takes about 12 years to complete, regardless of whether the center of that orbit is the earth (according to Ptolemy) or the sun (according to Copernicus).

The point about Saturn is less clear-cut and slightly different. Again with the telescope, Galileo had observed what appeared to him to be two bodies near this planet, although they did not seem to move relative to the planet. Relative to the earth, the system consisting of Saturn and its two "companions" appeared to change its position in various ways. Galileo had observed what today we know to be Saturn's rings, whose thickness is so small that they are invisible when looked edgewise from the earth, but whose width is so large that they are quite striking when oriented at their maximum inclination to our line of sight. However, he erroneously interpreted them to be bodies (like satellites), but fixed (unlike revolving satellites). On the other hand, in this passage, Galileo is expressing the correct intuition that Saturn undergoes axial rotation, with a period which is much shorter than the 30 years of its orbital revolution;[23] presumably he postulated this hypothesis in order to explain the pattern, sequence, and variations of appearance of the Saturn system. The key point here is that Saturn rotates on its axis while it revolves in its orbit, and so it moves with two natural motions.

The point about the sun is also different, but for a different reason. From the telescopic observation of the sun, Galileo had discovered the existence of sunspots and detected that they can be seen to move across the solar disk; and from this he had demonstrated that the sun undergoes axial rotation with a period of about one month. Now, for the case of the sun, Galileo obviously does not argue, and could not argue, that since it has the annual motion, then it too must have two natural motions; for it is not he, but his opponents, who attribute annual motion to the sun. However, what this means is that his opponents would be obliged to make that argument. In short, given the monthly axial rotation of the sun (provable from sunspots), the anti-Copernicans are committed to saying that it has at least two natural motions, and hence to reject the principle of the impossibility of multiple annual motions.

The way to question all this astronomical evidence is suggested by Simplicio: namely, to question the reliability of the telescope on which they depend. But Galileo, through Sagredo, dismisses this possibility. And there is some justice in this dismissal, for although that possibility was a serious one when the telescope was first developed in 1609 and for a few years afterwards, by the time of the publication of the *Dialogue* (1632) the reliability and legitimacy of the telescope had been established.

In any case, for the benefit of a skeptical opponent of the telescope, Galileo mentions a fourth piece of evidence. This additional evidence is available without the telescope and relates to magnetism and lodestones; it had been elaborated by William Gilbert, in *De magnete* (1600). Galileo discusses it at length in the Third Day (IIIB4, DML 461–81), but here he briefly mentions the essential point. That is, as heavy bodies, magnets have a natural motion of downward fall. However, they also have two characteristic motions: an horizontal motion that makes them seek a north–south orientation; and they also have a tendency to move along a vertical plane in order to point at an angle off the horizontal, by some amount depending on their location on the earth's surface. Thus, magnets have three natural motions.

In sum, the argument from the impossibility of multiple natural motions was an attempt by Chiaramonti to show that the motions attributed to the earth by Copernicus are metaphysically impossible. These metaphysical reasons involve the concepts of cause and effect and the notion of a simple body. In his criticism, Galileo argues that there are no good reasons for accepting the principle that simple bodies cannot have multiple natural motions, and there are some good empirical reasons for rejecting it (stemming from recent discoveries about the motions of heavenly bodies and the motion of magnetized bodies).

IIC5. LUMINOSITY AND MOBILITY (DML 307–19)

The Second Day ends with a section that briefly criticizes several of Chiaramonti's other objections to the earth's motion. With one exception, to be elaborated presently, these discussions mostly reiterate earlier points and do not add anything significant to the thread of Galileo's main argument. Thus they can be dealt with very briefly here.

For example, one of Chiaramonti's arguments (DML 307–8)[24] grounded the rejection of the earth's motion on the premise that substances with different natures have different motions; the difficulty is supposedly that on a moving earth such different elements as earth, water, and air would all be moving with the three (or rather two) motions which the Copernicans attribute to the terrestrial globe. Galileo replies that different substances, like water and air, need not have different motions completely, but only to the extent that their different natures can be inferred from some partial differences in motion or from some other behavior; hence, the diurnal and annual motions shared by water and air on a moving earth would not conflict with their dissimilar natures. The main issue here is almost identical to the one that emerged in Locher's argument from the inexplicability of terrestrial rotation (IIC2).

Another example (DML 310–12)[25] was Chiaramonti's version of the argument from the earth–heaven dichotomy. That is, if the earth moves, it would be a heavenly body, located in the heavenly region between the planets Venus and Mars; but the earth is a body full of change, corruption, and impurities, whereas the heavenly bodies are eternal, perfect, and pure; therefore, the earth cannot move. Galileo's answer consists of pointing out that this argument presupposes the earth–heaven dichotomy, and referring the reader to his earlier critique. In fact, the First Day argued that this dichotomy is invalidly justified by the Aristotelians, and is demonstrably false in light of recent astronomical discoveries.

Third (DML 312–14),[26] Chiaramonti had criticized one of Kepler's arguments in support of Copernicanism. Kepler had compared two particular features of the Ptolemaic and Copernican systems: Ptolemy attributed an extremely great speed of diurnal motion to the fixed stars (in their geocentric revolution); and Copernicus attributed an extremely large size to the universe, as measured by the great distance to the fixed stars (from the solar system). Kepler had argued that it was more reasonable for Copernicus to postulate the great stellar distances he did than for Ptolemy to postulate the great stellar speeds he did. Here, Galileo defends Kepler from Chiaramonti's criticism.

Fourth (DML 312–17),[27] Chiaramonti had objected that motion causes tiring, as shown by the observation of animal behavior; thus, the earth's motion would eventually stop, and hence it could not last forever. Galileo replies that it is not true that motion causes tiring, because the cause of animal tiring is the use of parts to move the whole; moreover, the earth's motion would be natural, whereas animal motion is violent; and even if motion caused tiring in the relevant sense, the earth would not tire any more than the *primum mobile* or the celestial sphere would get tired according to the geostatic system. The issues here echo those in the discussions of animal motion found in the last section (IIC4) and of conservation of motion in the corresponding section (IIB5).

Finally, we come to the more significant discussion mentioned earlier. A new anti-Copernican objection is stated and criticized, and the significance is that Galileo's criticism is not only destructive of a particular reason for rejecting the earth's motion, but also positive, insofar as it constitutes a particular reason for accepting the geokinetic thesis.

This particular argument by Chiaramonti attempted to refute Copernicanism based on the premise that substances of the same kind have correspondingly similar behavior with regard to motion and rest (DML 309).[28] He thought this premise was simply an empirical claim based on observation. Then he pointed out that Copernicanism contradicts this premise insofar as it holds that six of the heavenly bodies—the planets—move circularly around the center of the universe, whereas the sun and the fixed stars do not.

In this argument, Chiaramonti is assuming that the planets, moon, sun, and fixed stars are bodies of the same nature. If challenged to justify this assumption, he probably would refer to their sharing such properties as being composed of the element aether, lacking any physical changes (except motion, i.e., change of location), and being intrinsically luminous. This argument complements the first one described above in this section: there the major premise was that bodies of different kinds have different behavior with regard to motion, whereas here the major premise is that bodies of the same kind have the same kind of motion; these claims are not equivalent, but rather the converse of each other.

A version of this argument, explicitly focused on the alleged correlation between luminosity and mobility, was a common Aristotelian objection against Copernicanism (DML 53–54).[29] It claimed explicitly that the earth cannot move because it is devoid of its own intrinsic light, and only bodies shining with their own light are in (natural circular) motion, as revealed by the observation of the sun, moon, planets, and fixed stars.

Galileo's criticism is incisive enough to deserve extended quotation:

> The form of the argument appears to me valid, but I believe that its content or its application is wrong; and as long as the author wants to persist in his undertaking, the result will run directly counter to his.
> The counterargument proceeds as follows. Among world bodies, there are six that perpetually move, and they are the six planets. Of the others (that is, the earth, the sun, and the fixed stars), the question is which move and which stand still; for it is necessary that if the earth

stands still then the sun and the fixed stars move, and it is also possible that if the earth moves then the sun and the fixed stars are motionless. Not knowing the fact of the matter, we are trying to determine to which ones it is more convenient to attribute motion, and to which ones rest.

Natural reason tells us that we should consider motion to belong to a body whose kind or essence corresponds better to that of bodies which undoubtedly move, and rest to a body which is more dissimilar from them. And since perpetual rest and perpetual motion are extremely different properties, it is manifest that the nature of a body which is always in motion should be extremely different from the nature of one which is always at rest. Thus, given that we are uncertain about who is in motion and who is at rest, we are searching for some other relevant condition that would enable us to determine which body is more similar to those which are certainly in motion: whether it's the earth, or the sun and the fixed stars.

But behold how nature, as if to favor our need and desire, provides us with two striking conditions that differ no less than motion and rest. They are light and darkness, that is, being luminous by nature, and being dark and devoid of any light. Thus, bodies endowed with an internal and eternal brilliance are extremely different in essence from bodies devoid of any light. Now, the earth is devoid of light; the sun is extremely brilliant in itself, as also are the fixed stars; and the six moving planets totally lack any light, like the earth. Therefore, their essence corresponds to that of the earth, and differs from that of the sun and fixed stars; hence, the earth is in motion, and the sun and stellar sphere are motionless.

(FAV 291; cf. DML 309–10)

Here, Galileo explicitly says that this is not merely a negative criticism of Chiaramonti's argument from the similarity of motions of similar substances, but also a positive argument for the earth's motion. It undermines Chiaramonti's argument by refuting his claim that Copernicanism contradicts the principle that bodies with the same nature have the same kind of motion. But to deny that Copernicanism contradicts this principle is to assert that Copernicanism is consistent with it, and implicitly that the Ptolemaic system contradicts it. Thus, once we accept this principle, we have a reason for accepting Copernicanism and rejecting the Ptolemaic system.

As Simplicio is quick to point out, Galileo's counterargument depends on the claim that the planets do not give off their own light, and the Aristotelians would deny this claim and assert that they are intrinsically luminous. The discussion would then move to that issue. But on that issue they were wrong, and the new telescopic discoveries could conclusively prove their error. And so Galileo's criticism undermines the argument from the correlation between luminosity and mobility, by refuting this alleged correlation, and indeed reversing it and claiming that the correlation is instead between mobility and nonluminosity.

However, it would be another error to claim that Galileo's argument establishes conclusively that the earth moves. As the language used in this passage to express this argument makes clear, this is an inductive argument based on empirical and approximate correlations, and its force is merely probable. He is even more explicit about this judgment in another statement he gives of this argument, in the "Reply to Ingoli" (Galilei 1890–1909, vol. 6: 559–61; Finocchiaro 1989: 96–97).

NOTES

1 See also DML 144; Aristotle, *On the Heavens*, bk 2, ch. 14, 296a28–34.

2 I have changed the "Prime Mobile" in FIN 143 to *"primum mobile."* See also DML 158; Aristotle, *On the Heavens*, bk 2, ch. 14, 296a34-b6.

3 Cf. Besomi-Helbing 306–7, Finocchiaro 2010a, Strauss 521 n. 38 (= Sexl-Meyenn 521 n. 38).

4 Galilei 1890–1909, vol. 5: 545; in Finocchiaro 1989: 184. Cf. Conti 1990: 230–31; Feyerabend 1988: 67, 77 n. 20; Finocchiaro 2010a; Koyré 1966: 226–28, 1968: 13, 1978: 166–67.

5 Cf. Aristotle, *On the Heavens*, bk 3, ch. 2, 301b22–30; *Physics*, bk 8, ch. 10, 266b27–267a21.

6 Respectively: FAV 192.24; Webbe 1635, folio 118v; Salusbury 146 and Santillana 180; DML 193; Shea-Davie 249.

7 "Absolute," not of course in the sense of post-Einstein physics, but relative to some coherent notions of common sense. That is, we are not claiming that such a speed is measured to be the same by all observers regardless of their motion, as is the case for the speed of light; rather we are talking about speed as one might (imagine to) observe it in a reference frame centered at the sun.

8 Cf. Gapaillard (1992, 1993: 195–202), Graney (2011a, 2011b).

9 Koyré 1966: 215–38; DML 147–48, 195–96, 209. Cf. Finocchiaro 1980: 208, 240, 399–403, 412 n. 13.

10 I have corrected Drake's "lively" to "animate"; cf. FAV 209, Shea-Davie 267–68, Besomi-Helbing 482.

11 DML 218, FIN 171, FAV 214, DCA 188. Cf. Ptolemy, *The Almagest*, bk 1, ch. 7 (1952: 10–12); Copernicus, *On the Revolutions*, bk 1, ch. 7 (1992: 15); Strauss 519–20 n. 30 (= Sexl-Meyenn 519–20 n. 30); Santillana 146 n. 33; Hill 1984: 110–15; Palmieri 2008.

12 Cf. Hill 1984, Finocchiaro (2003, 2010b: 97–120), Palmieri 2008, Heilbron 2010: 272–74.

13 Unless one took into account the irregularity of the earth's surface; for then a body that had reached orbital velocity could be struck later by a mountain moving at a higher velocity due to the increased terrestrial rotation, and so the body would acquire additional velocity; eventually, such additions might increase its velocity to that required for escape extrusion. I thank Albert DiCanzio for this refinement; cf. also DiCanzio 1996: 144–45, 171–73, 353–54.

14 See Chalmers and Nicholas 1983: 323–28, Gaukroger 1978: 193–95, Pagnini vol. 2: 415 n.

15 Cf. Locher 1614: 28–31, Koyré 1955: 330–33, Gapaillard (1990–91, 1993: 203–15), DiCanzio 1996: 170–71, Besomi-Helbing 534–39, Heilbron 2010: 275–76.

16 For another possible explanation, which conjectures a typographical error, see Heilbron 2010: 436 n. 93.

17 Cf. also Locher 1614: 31–34, Besomi-Helbing 566–67, Shea-Davie 297–98.

18 Here my translation amounts essentially to making numerous emendations to Drake's.

19 This estimate is actually about 25 times smaller than the true modern value; see Pagnini vol. 2: 464 n.

20 Cf. also Chiaramonti 1628: 472–73, Besomi-Helbing 586.

21 This is my translation, obtained by starting with Drake's and emending it in various ways to improve accuracy. Cf. Galilei 1632: 250–51, Chiaramonti 1628: 473–75, Besomi-Helbing 598–99, Beltrán 223.

22 Here this point is just briefly stated, but later (DML 462–63) Galileo gives a more elaborate explanation. It also comes up in section IIIC3 below, and that is where I will explain it at greater length.

23 According to modern astronomy, Saturn's period of axial rotation is 10.5 hours, and its period of orbital revolution 29.5 years. Saturn's rings were effectively discovered by Huygens in 1655–56.

24 Cf. Chiaramonti 1628: 477, Besomi-Helbing 610.

25 Cf. Chiaramonti 1628: 481–82, Besomi-Helbing 614–15.

26 Cf. Chiaramonti 1628: 483–84, Besomi-Helbing 618–19.

27 Cf. Chiaramonti 1628: 482–83, Besomi-Helbing 617.

28 Cf. Chiaramonti 1628: 481, Besomi-Helbing 612–13.

29 Cf. Besomi-Helbing 228, Bucciantini and Camerota 2009: 185–86.

6

DAY III

EARTH'S ANNUAL HELIOCENTRIC REVOLUTION

IIIA. EVIDENCE FROM THE 1572 NOVA (DML 321–70)

In his book *On the Three New Stars*, Chiaramonti had argued that the nova of 1572 was located at a distance less than that of the earth to the moon. The significance of this conclusion was that it preserved the earth–heaven dichotomy, since the nova could then be regarded as another phenomenon in the terrestrial region, rather than a change in the heavenly region. Chiaramonti's conclusion was based on an analysis of the observational data of about a dozen astronomers. He compared and contrasted the nova's apparent change in position and elevation resulting from the different locations of the observers on the earth's surface, with the corresponding change in the apparent elevation of the celestial pole. He found considerable discrepancy in the distances computed from the various sets of data, but he thought that, by and large, the computations pointed toward a sublunary location for the nova.

Galileo begins the Third Day with a critique of this argument. The main thing he does is to re-analyze the data used by Chiaramonti, using the same method of comparing and contrasting the change in apparent elevation of the nova and of the pole due to the change of the observer's location. Galileo points out that the data yield physically impossible distances in most cases, and widely different distances for the rest. The physically impossible distances are those implied by negative parallax differences; the other distances vary from 1/48 to 716 earth radii. Thus, the proper procedure to follow is either to reject all the data as unreliable, or else to correct them appropriately. Chiaramonti failed to do either, and this was his main error.

One way to correct the observational data would be from the viewpoint of the qualitative question whether the nova is sublunary or superlunary. The data could then be subdivided into those yielding sublunary distances and those yielding superlunary distances. One would then try to harmonize the sublunary set of data on the one hand, and the superlunary data on the other, by making appropriate corrections to the data. When this is done, it turns out that the corrections needed to reconcile the superlunary set of data are much smaller than those needed for the sublunary data. From this one could conclude that the nova was probably superlunary.

Another way to correct the observational data would be from the perspective of the range of distances suggested by most observations. Then only the data yielding impossible distances would be changed. Now, almost all the data yield impossibly great distances, but small corrections to them yield possible but great distances (far beyond the moon). Thus, we may conclude that the great majority of the data suggest a superlunary distance.

Galileo also argues that the same conclusion may be obtained by using a method not exploited by Chiaramonti. That is, the most reliable data are the measurements of the angular distance between the nova and the celestial pole, or between the nova and some neighboring star. Such measurements are relatively easy to make, and computation from them yields superlunary distances.

Thus, Galileo can add the evidence of the 1572 nova to his argument presented in the First Day, disconfirming the earth–heaven dichotomy and confirming the thesis that the earthly and heavenly regions are similar from the point of view of undergoing change. This is no proof, or even direct argument, for the earth's motion, but it represents a further weakening of one argument against it. More generally, Galileo is suggesting that quantitative astronomical data are problematic and great care must be exercised in drawing conclusions from them.

IIIB1. MARS, VENUS, AND MOON (DML 381–95)

The critique of the evidence from the 1572 nova is immediately followed by the presentation of a pro-Copernican argument based on the heliocentrism of planetary motion. However, this argument utilizes observational data not available before the telescope, which had enabled Galileo to answer some very powerful objections to the earth's annual motion. Thus, before presenting that pro-Copernican argument, it is better to discuss these objections first.

These objections were arguments based on the behavior and appearance of the moon and the planets Mars and Venus, as observed with the naked eye. Galileo thought that they were so strong that, without the telescopic evidence, he himself would have been unable to reject the geostatic thesis and pursue the geokinetic hypothesis. In this regard, he contrasts himself with Aristarchus and Copernicus who, based on more intellectual and theoretical arguments, had managed to convince themselves of the earth's motion, even though it contradicted the incontrovertible evidence of astronomical observation.

In light of these considerations, it is not surprising that Galileo gives statements of these objections using inspired language such as the following:

> I have already explained to you the structure of the Copernican system. Against its truth the first extremely fierce assault comes from Mars itself: if it were true that its distance from the earth varies such that the farthest minus the closest distance equals twice the distance from the earth to the sun, it would be necessary that when it is closest to us its disk should appear more than sixty times larger than when it is farthest; however, this variation of apparent size is not perceived; instead, at opposition to the sun, when it is close to the earth, it

appears barely four or five times larger than near conjunction, when it is hidden behind the sun's rays. Another and greater difficulty is due to Venus: if (as Copernicus claims) it should turn around the sun and be sometimes beyond the sun and sometimes in between, and if it should recede from us and approach us by a difference equal to the diameter of the circle it describes, then when it is positioned between us and the sun and is closest to us its disk would appear almost forty times larger than when it is positioned beyond the sun and is near its other conjunction; however, the difference is almost imperceptible. To this we should add another difficulty: it seems reasonable that the body of Venus is inherently dark and shines only because of the sun's illumination, like the moon; if this is so, then when positioned between us and the sun it should appear sickle-shaped, as the moon does when it is likewise near the sun; but this phenomenon is not observed in Venus ... Furthermore, there is a feature that alters the order in such a way as to render it unlikely and false: all planets together with the earth move around the sun, which is at the center of their revolutions; only the moon perturbs this order, by performing its proper motion around the earth; and then it, the earth, and the whole elemental sphere all together move around the sun in one year.

(FIN 235–37; cf. DML 388)

Here we have four items of counterevidence, one regarding Mars, two involving Venus, and one about the moon. The observational data were uncontested, being easily obtained with naked-eye observations. And the observational consequences of the Copernican hypothesis were also unquestionable: in the Copernican system, the earth is the third planet revolving around the sun and completes its orbit in one year; Venus is the second planet and completes its smaller orbit in about 7 1/2 months; and Mars is the fourth planet and completes its larger orbit in about 2 years. As both the earth and Mars revolve around the sun at different rates, there would be times when the two planets would be on the same side from the sun and thus relatively close to each other, and other times when they would be on opposite sides of the sun and thus relatively far from each other. Similar changes would happen to the distance between Venus and the earth. Given the estimates of distances commonly shared at that time, the distance between the earth and Mars would change by a factor of about 8, and that between the earth and Venus by about 6. Thus, their apparent diameters would vary, respectively, by the same factors. But the apparent size, as measured by the size of their disk, would vary by the square of these numbers, since areas vary as the square of linear dimensions; and here Galileo is rounding off the square of 8 (= 64) to 60, and the square of 6 (= 36) to 40. However, observation revealed no significant variation in the size of Venus, and only a variation of 4 or 5 times in the case of Mars.

Moreover, for the case of Venus there would an additional observational consequence. It would have to exhibit phases, with its shape changing like the moon, except that the period would be different (several months instead of one). That is, as Venus and the earth revolve around the sun at different rates, there would be times when Venus would be located between the sun and the earth, and thus it should be seen in a crescent shape; at other times, Venus would be on the opposite side of the sun relative to the earth, and so it should appear as a full round disk; and at times in between its shape would be changing from crescent to full, and from full to crescent, going through a phase of half visibility. However, observation revealed no such phases. As Galileo makes clear in the quoted passage, this objection assumes that Venus is not intrinsically luminous, but rather dark and opaque. However, this assumption was accepted by the Copernicans, and so they could not criticize this objection by questioning this assumption.

Finally, what is the problem stemming from the moon? Galileo says that in the Copernican system it "perturbs" the order of the planets and their heliocentric revolutions: Mercury, Venus, Earth, Mars, Jupiter, and Saturn. This means partly that for Copernicus the moon is not exactly a planet: it is in part a planet insofar as it moves relative to the fixed stars in an orbit that is completed in a fixed period (one month); but it revolves directly around the earth, rather than around the sun; it does revolve indirectly around the sun, but only because somehow it is carried by the earth as the latter revolves around the sun. The last point embodies the real difficulty: according to Copernicanism, the moon accompanies the earth in its motion around the sun, even though there is no known force tying the two bodies together.

Galileo's refutation of these four objections is based on the observations made possible by the telescope. In each case, a key premise of the anti-Copernican argument is false. Telescopic observation reveals that the apparent disk of Mars varies in size by a factor of about 60; that the disk of Venus changes in apparent size by a factor of about 40; and that this same planet exhibits phases.

Regarding the lunar-orbit objection, the answer is more complicated. The relevant telescopic observation is that the planet Jupiter has four moons revolving around it at different distances and in deferent periods. Now, both sides of the controversy agreed that Jupiter is a planet revolving in an orbit that encompasses both the earth and the sun and is completed in about 12 years. Therefore, Jupiter's satellites show that it is possible in nature for one body to revolve around another, while the latter revolves in an orbit around some other point. If the earth has an annual motion, then the moon's revolution around it would be simply an instance of the same natural phenomenon. That is, the moon would not disturb the Copernican order of the planetary motions, any more than the demonstrably real moons of Jupiter disturb the commonly agreed-upon order of the planetary motions.

This Galilean criticism could be opposed, and was indeed opposed, by questioning the instrument used to make such novel observations. Indeed, if one did not accept the legitimacy or reliability of the telescope, then one did not have to accept these discoveries. The issue has arisen before (IIC4) in connection with Chiaramonti's objection from the impossibility of multiple natural motions, which Galileo refuted based on telescopic evidence about Jupiter's satellites, Saturn's axial rotation, and sunspot motion. As before, the key point is that for a few years after the telescope was introduced in 1609 there were indeed proper questions about its methodological legitimacy and empirical reliability, but that by the time the *Dialogue* was published (1632) this controversy had been essentially settled.

A more relevant issue is raised by Sagredo when he shrewdly asks, "But Venus and Mars are not objects that are invisible because of their distance or small size; indeed we perceive them with our simple natural vision. So why do we not distinguish the variations in their size and shape?" (FIN 238; cf. DML 390). For even if we agree that the telescope reliably magnifies objects by as much as 30 or 40 times, and thus renders visible things that are invisible to the naked eye, why should we not with the naked eye see the significant variation in the apparent size of Mars and Venus and the phases of the latter? To strengthen his criticism, Galileo must give an explanation of why the naked eye does not reveal these particular features, which the telescope reveals.

The explanation is this.[1] Although the eye is a natural organ, it is itself a physical instrument whose operation involves physical processes, consisting primarily of an interaction between it and the light coming from the visible object. In normal observation of small and brilliant objects, when their light enters the naked eye, and before it strikes the retina and its signals reach the brain, it undergoes various processes such as refraction through the fluids of the eye, and reflection by the eyelids; thus, objects appear surrounded by rays, as if encased in a head of hair. This irradiation makes their apparent size seem larger than it really is.

This enlargement is partly an increasing function of how small the true and unadorned apparent size is. For example, suppose we are observing a body whose true apparent size is two seconds of arc in diameter, and that the naked eye adds a head of hair which is four seconds wide on each side; then the enlargement would be 25 times, because the disk has changed from a diameter of 2 to a diameter of 10, and apparent size is measured by the area, which varies as the square of the diameter (from 4 to 100). But suppose the object under observation has a true apparent size of just one second in diameter; now, when the eye adds four seconds on each side, we go from a diameter of 1 to a diameter of 9, and hence from an area of 1 to an area of 81, and hence to an 81-fold increase in apparent size.

Now, when we observe small brilliant objects with a telescope, the irradiation effect does not occur. Similarly, it does not occur even if we observe them through a small hole in a piece of paper, or through a small slit made with our hand or fingers. Thus, planets and stars actually appear smaller when seen with a telescope. In particular, without the irradiation contributed by the naked eye, Mars and Venus appear to undergo the requisite variation in apparent size. In normal naked-eye observation, these variations are masked by the enlargement due to the irradiation. Thus, for example, when Mars increases in true apparent size by 60 times, the irradiation effect magnifies the smaller size by a larger relative amount, whereas it magnifies the larger size by a much smaller relative amount; the result is that what's visible with the naked eye is a much smaller change, such as the 4 or 5 times actually seen.

For the case of Venus, the situation is complicated by other factors. First, the phenomenon of the phases interferes with, and counterbalances, the phenomenon of the change in apparent size. That is, Venus is closest to the earth when it is between us and the sun, and thus with a telescope it appears both crescent and much larger (in the sense of apparent diameter); but the greater light due to the greater diameter is counterbalanced by the smaller amount of visible surface due to its crescent shape. Conversely, Venus is farthest when it is on the opposite side of the sun from where the earth is; and thus with a telescope it appears both as a full disk and much smaller; but the smaller size of the disk counterbalances the fact that the whole disk is visible. With the naked eye, these two situations are not distinguishable, partly because of the masking by the irradiation effect, and partly because of the counterbalancing of the phases vis-à-vis the changed sizes.

Finally, the irradiation effect is also an increasing function of the brilliance of the visible object. This means that when Mars is closest to the earth, and thus should appear larger, it is also farthest from the sun, and so the light it receives from the sun and sends to us is less intense; thus, in naked-eye observation, the irradiation effect and the enlargement are smaller. However, when Mars is farthest from us and should appear smaller, it is closer to the sun and more brilliant, and so when observed with the naked eye the irradiation effect and enlargement are greater. Regarding Venus, when it should appear larger, it is in its crescent phase, and thus the light sent to us is fainter and more indirect; whereas when it should appear smaller, it is in its full phase, and so the light we see is more direct and brighter; this again contributes to the larger true apparent size being enlarged less, to the smaller true apparent size being magnified more, and to the naked eye seeing both as about equal.

This is how it happens that with the naked eye we see things that are not there, but the telescope makes us see them as they really are, so to speak. But here Galileo is talking about the observation of apparently small brilliant objects, such as the heavenly bodies (other than the sun and moon). Nevertheless, this explains how and why the naked eye is misled in the observation of Mars and Venus, and how and why the telescope enables us to see what we should be able to see if the earth had the annual motion hypothesized by Copernicus. However, so far this merely refutes the anti-Copernican arguments from the apparent size of Mars and Venus, and from the phases of Venus; it does not prove the earth's motion. On the other hand, the telescopic evidence mentioned so far can be used to argue in favor of that hypothesis. This is done by Galileo in the passage which I discuss next, although, as mentioned earlier, he presents it immediately before the present section.

IIIC1. HELIOCENTRISM OF PLANETARY REVOLUTIONS (DML 370–80)

The heliocentrism of planetary revolutions is a handy label for the thesis that the planets Mercury, Venus, Mars, Jupiter, and Saturn revolve in orbits whose center is the sun (rather than the earth). This particular thesis does not regard the earth as one of the planets, and so does not assert that the earth too revolves around the sun. The thesis is referring only to the five bodies which both sides of the controversy agreed were planets, moving westward in orbits of different sizes, and taking different periods of time to complete them. The earth's annual heliocentric motion is a separate issue, which Galileo does also address in this section, but he does so only after he has established that the center of the orbits of these five planets is the sun.

In fact, this passage contains an argument with two main steps. The first is a subargument supporting the heliocentrism of planetary revolutions based on evidence that is partly old and partly new. The old evidence is easily obtained with naked-eye observation, and was known since ancient times, and was completely uncontroversial in Galileo's time. The new evidence relies on the telescope and stems from Galileo's own discoveries; although it had been controversial at first, by the time he wrote the *Dialogue* it, too, was relatively uncontroversial. Indeed, by this time, the heliocentrism of planetary motion was itself a thesis that was widely accepted and relatively uncontroversial. In particular, it provided the central tenet in Tycho's system, according to which the planets revolve around the sun, although the sun with the whole planetary system moves daily and annually around a motionless and central earth. Thus, what gives interest to this part of Galileo's argument is not the conclusion it reaches, but rather the clarity, incisiveness, and elegance of his presentation of the evidence.

The second step of the main argument is a subargument supporting the conclusion that the earth too revolves annually around the sun, based on the heliocentrism of planetary revolutions—the thesis previously established. This is the more controversial part of the argument; as we shall see, it is weak, but ingenious and plausible.

To prove the heliocentrism of planetary revolutions, Galileo reviews the evidence regarding each planet in turn. In presenting this evidence, he uses several semi-technical terms, which it is proper for us to explain at the outset. These terms are labels used to refer to the relative position of two heavenly bodies on the celestial sphere, in terms of the angular distance between them as measured by a terrestrial observer. Two bodies are said to be in *opposition* (to each other) when they appear on opposite sides of the celestial sphere, namely 180 degrees apart. They are said to be in *conjunction* when they appear next or near each other on the celestial sphere, namely at an angular distance close to zero. They are said to be in *quadrature* when they appear about 90 degrees apart on the celestial sphere.

The argument begins with Venus, regarding which the evidence is most detailed. It is easily observed and well known that Venus always appears located near the sun, the maximum angular separation being about half a quadrature; this suggests that its orbit is somehow associated with the sun. In particular, Venus is never seen in opposition to the sun; this proves that its orbit does not encompass the earth. Finally, the phases of Venus prove that its orbit encircles the sun: for when it appears large, crescent, and in conjunction, it is located between the earth and the sun; and when Venus appears smaller, full, and in conjunction, it is located beyond the sun, namely the sun is between the earth and Venus. Moreover, Venus's orbit cannot be entirely in the space between the earth and the sun, because in that case it would always appear crescent, and never full; and its orbit cannot be located entirely beyond the sun, because in that case it would always appear full or gibbous, and never crescent. Undoubtedly, therefore, Venus moves in an orbit around the sun that does not encompass the earth.

Next, Mercury behaves in similar manner. It always appears near the sun, and indeed its maximum separation from the sun is much smaller than that of Venus. Galileo does not say that Mercury too exhibits phases, since he had not observed them. However, he plausibly explained his failure to see Mercury's phases as follows: because of its closeness to the sun, it is clearly visible only when it reaches maximum separation, on either side of the sun; but at these two positions its distance from the earth is the same, and so no change in apparent size is detectable; also at these two positions, it should appear about half full, but its closeness to the sun makes it so brilliant as to generate a large irradiation effect which available telescopes cannot resolve. Thus, Galileo concludes that Mercury moves around the sun in an orbit smaller than that of Venus.

With regard to Mars, unlike Venus and Mercury, it is sometimes seen in opposition, and at such a location it appears very large and thus very close to the earth. But when Mars is in conjunction with the sun, it appears very small, and hence very far from the earth. Telescopic observation reveals the difference represents a change of about 60 times in apparent size, or 8 in distance. Moreover, at such a location it must be beyond the sun, and not between the earth and the sun, because it appears full and not crescent; indeed, Mars shows no phases, but always appears as a round disk. It follows that Mars moves in an orbit which encloses both the earth and the sun; which is definitely not centered on the earth; and which is likely to be centered on the sun.

Finally, both Jupiter and Saturn behave like Mars. Each is seen sometimes in opposition and sometimes in conjunction; thus, their orbits encompass the earth. Further, their orbits encompass the sun because in conjunction they appear full and not crescent. But these orbits are not centered at the earth, because they appear smaller in conjunction and larger in opposition. So they are likely centered on the sun. And since the changes in apparent size are larger for Jupiter and smaller for Saturn, the orbit of Jupiter is smaller than that of Saturn.

This is the subargument showing that these five planets revolve in orbits which are centered on the sun (not the earth) and which are of different sizes and with different periods. However, as previously mentioned, terrestrial motion, diurnal motion, and annual motion are nowhere in sight yet. All that has been demonstrated so far is something about the relative locations of planets, sun, and the earth, and about the revolutions of the planets. To address the issue of the earth's motion, Galileo argues as follows:

> Three things now remain to be assigned to the sun, earth, and stellar sphere: that is, rest, which appears to belong to the earth; the annual motion along the zodiac, which appears to belong to the sun; and the diurnal motion, which appears to belong to the stellar sphere and to be shared by all the rest of the universe except the earth. Since it is true that all the orbs of the planets (namely, Mercury, Venus, Mars, Jupiter, and Saturn) move around the sun as their center, it seems much more reasonable that rest belongs to the sun than to the earth, inasmuch as it is more reasonable that the center of moving spheres rather than any other point away from this center is motionless; therefore, leaving the state of rest for the sun, it is very appropriate to attribute the annual motion to the earth, which is located in the middle of moving parts; that is, between Venus and Mars, the first of which completes its revolution in nine months, and the second in two years. If this is so, then it follows as a necessary consequence that the diurnal motion also belongs to the earth; for if the sun were standing still and the earth did not rotate upon itself but only had the annual motion around the sun, then the cycle of night and day would be exactly one year long; that is, we would have six months of daylight and six months of night, as we have stated other times. So you see how appropriately the extremely rapid motion of twenty-four hours is taken away from the universe, and how the fixed stars (which are so many suns) enjoy perpetual rest like our sun. Notice also how elegant

this first sketch is for the purpose of explaining why such significant phenomena appear in the heavenly bodies.

(FIN 232–33; cf. DML 379–80)

This plausible, but nonconclusive, argument advances three main reasons for the earth's motion: it is more fitting to have the center (the sun) rather than a point off center (the earth) be motionless; the earth is positioned between two other bodies (Venus and Mars) which perform orbital revolutions; and the period of the earth's orbital revolution (one year) is intermediate between the periods of Venus and Mars, just as the size of its orbit is intermediate between the sizes of theirs. Salviati's talk of "elegance" in the last sentence of this passage, and Sagredo's talk of "simplicity" in his next response are clues that this is, and is meant to be, a simplicity argument. This content and terminology also suggest that an aesthetic value may be involved, perhaps an aspect of the concept of simplicity, or a distinct principle. However, a conclusive point is that if the earth has an annual motion, then it must have the diurnal motion, otherwise the cycle of night and day would last one year.

IIIC2. RETROGRADE PLANETARY MOTIONS (DML 396–400)

Retrograde planetary motion refers to the phenomenon that, in the course of their eastward orbital revolutions, planets are periodically seen to slow down, stop, reverse course, briefly move westward ("backward") before resuming their regular eastward motion. Moreover, during retrogression planets appear brighter, as if they were nearer the earth than at other times. This phenomenon can be observed with the naked eye, and it occurs only for the five planets mentioned earlier (Mercury, Venus, Mars, Jupiter, and Saturn).

Proponents of the geostatic view from Aristotle, through Ptolemy, to Tycho had attempted to explain it by attributing to these bodies an additional motion. Copernicus, on the other hand, held that retrograde motion could be explained on the basis of the earth's annual motion, and that such a geokinetic explanation is better than the geostatic one. Galileo followed Copernicus in this regard, and in this section he gives an incisive presentation of the Copernican explanation, and some reasons for its superiority over the alternative.

Although Galileo summarizes in one sentence the key feature of the geostatic explanation of planetary retrogression, he does not bother to elaborate it, since at that time it was well known. Thus, will be useful for us here to give a brief account of that alternative explanation. This can be done best by reference to Figure 6.1.

A deferent was defined as a geocentric circle whose circumference (ABCD) rotated around the earth (E). An epicycle was defined as a circle (FGHI) whose center (A) lay on the circumference of the deferent, and whose circumference rotated in the same direction as the deferent. The planet was located on the circumference of the epicycle. Thus, when the rotation of the epicycle carried the planet on the far side (F) of the epicycle from the earth, its distance was the sum of the radii of the deferent and the epicycle; whereas when the epiclic rotation carried the planet on the near side (H) of the epicycle from the earth, its

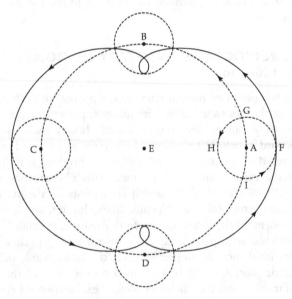

Figure 6.1

distance was the difference between the two radii. Thus, in its geocentric revolution, the distance of the planet from the earth changed by an amount equal to the diameter of the epicycle. This difference accounted for the variation in brightness.

Moreover, the planet's motion was the result of its motion along the epicycle and the motion of the center of the epicycle along the deferent. Thus, when the planet was on the far side (F), its speed was the sum of the deferent speed and the epicycle speed, and hence was faster than its average speed. But when the planet was on the near side (H) of the deferent, its speed was the difference between the two; in this case, if the epicycle speed was greater than that of the deferent, the planet appeared to move backwards (clockwise or westward). Retrograde planetary motion then resulted.

For each planet, the relative sizes of deferent and epicycle and their relative rates of rotation could be adjusted so that their combination yielded mathematically the observed details about retrogression and changes in brightness and speed. For example, if the planet was observed to retrogress twice while revolving through its complete orbit once, then the epicycle was assumed to rotate twice as fast as the deferent; this yielded a path which in reality was looped; but from the earth (E) the loop was not seen, and instead the planet would appear brighter and retrogressing near B and D.

This framework of deferents and epicycles was a very powerful instrument for the analysis of planetary motion. There was much more that an astronomer could do besides adjusting the relative sizes and speeds of a deferent and its epicycle. For example, one could add a second epicycle on the first epicycle; one could make the center of the deferent different from the center of the earth, in which case the deferent was called an *eccentric*; and one could even make the center of the deferent move in some way, perhaps in a small circle around the earth's center. For many centuries before Copernicus, such calculations, adjustments, and refinements involving deferents, epicycles, and eccentrics constituted the primary theoretical and mathematical task of planetary astronomy. This enhanced the power of the theory, but it also rendered the whole system increasingly complicated.

Such complications could be largely avoided in the Copernican system. Instead, retrograde motion is a direct consequence of the relative motion between the earth and the other planets. The key point is that when the earth and another planet reach points in their orbits which are on the same side of the sun and are thus at the minimum distance from one another, their different speeds make the other planet seem to move "backward" (westward) as seen from the earth.

This is more easily explained for the case of the superior planets, namely planets whose orbit is larger than the earth's (Mars, Jupiter, Saturn). Their retrogression happens when they are in opposition to the sun; the earth's faster eastward motion leaves behind the other planet, which thus appears to move westward relative to the fixed stars. Let us refer to Figure 6.2 (DML 399), where O is the sun, the smaller circle is the earth's orbit, the larger circle is the orbit of a superior planet, and the curved line on the extreme right is a portion of the celestial sphere. Now, while the earth moves through points B, C, D, E, F, G, H, I, K, L, and M along the smaller orbit, the superior planet moves along its bigger orbit through a corresponding set of points comprising a shorter distance due to its slower speed. But the apparent position of the planet against the background of the fixed stars changes in the order P,

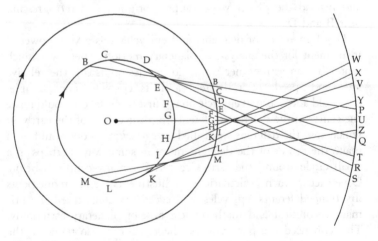

Figure 6.2

Q, R, S, T, U, V, W, X, Y, and Z. And the motion from P to S is eastward, or "direct"; from S to W, it is westward, or retrograde; and from W to Z, it is eastward, or "direct," once again.

For the case of an inferior planet (Venus and Mercury), the phenomenon occurs when it is in conjunction with the sun. The earth's slower motion enables the other planet to overtake the earth relative to the fixed stars, thus generating the appearance that the planet is moving in a direction opposite to that of the sun. Since the latter always appears to move eastward relative to the fixed stars, the planet appears to move westward.

Galileo claims that the geokinetic explanation is better than the epicyclic one for several reasons. First, even if the two accounts were approximately equivalent from the instrumentalist point of view of mathematical calculation and observational prediction, the geokinetic explanation is superior from the realist point of view of philosophical understanding and description of what is really going on in physical reality. Second, the epicyclic explanation is *ad hoc*: that is, the size and speed of the epicycles are adjusted "one by one for each planet" (FAV 370; cf. DML 397), so as to yield the observed features of their motions; whereas the Copernican explanation is more systemic, because by using the otherwise known features of planetary motions, all retrograde motions can be explained by the earth's annual motion. Third, the geokinetic explanation can avoid various absurdities of the geostatic account, such as having to make the epicycle of Mars so large that this planet intersects the orbit of the sun and gets closer to the earth than the sun does. Finally, the Copernican explanation has greater "elegance and simplicity" (FAV 372; cf. DML 400).

Galileo is aware that these are methodological judgments which are not necessarily shared by everyone, and that therefore his geokinetic argument from retrograde planetary motion is, at best, only a probable one. Nevertheless, it can be added to the list of reasons to prefer the Copernican to the geostatic world view.

IIIC3. PATHS OF SUNSPOTS (DML 400–414)

The discussion now proceeds to a topic which, unlike retrograde planetary motion, involves empirical facts that were little known, had been recently discovered, relied on the use of the telescope, and were highly controversial in several ways; that is, the topic of the annual variation in the paths of sunspots.[2] However, this phenomenon provides the basis for a pro-Copernican argument whose logical structure is similar to the argument from retrograde planetary motion: devising alternative theoretical explanations of the observational facts, and suggesting that one (the geokinetic) explanation is better than the other (the geostatic one).

The controversy was primarily a dispute between Galileo and the Jesuit astronomer Scheiner concerning the priority of discovery of sunspots, and their precise observational features, theoretical interpretation, and cosmological significance. The controversy had started in 1612, when the two exchanged some correspondence and wrote several works, including Galileo's *History and Demonstrations Concerning Sunspots* (1613). At that time, there were three principal issues: who had been the first to observe sunspots; whether sunspots are phenomena occurring on the surface of the sun, or swarms of previously undetected planets revolving around it, which darken parts of the solar disk when some of them get into the line of sight of a terrestrial observer; and depending on which of these alternatives is correct, and given that sunspots appear to move across the solar disk and exhibit corresponding changes of size and speed, whether the sun undergoes an axial rotation with a period of about a month.

Concerning priority, Galileo and Scheiner probably observed sunspots independently of each other and at about the same time; indeed, so did other astronomers. Regarding the location of sunspots, Galileo was right that they occur on the solar body and are carried along as the sun rotates on its axis once a month. Eventually, Scheiner accepted this Galilean thesis. In fact, Scheiner refined and systematized his observation and study of sunspots, and in 1630 he published a monumental work (*Rosa Ursina*) full of observational details and theoretical discussions, including a chapter of venomous criticism of Galileo. The main discovery detailed in this book was a phenomenon which neither he nor Galileo had detected in 1612–13: the sun's axis of monthly rotation is inclined to the plane of the ecliptic by about 7 degrees; as they move across the solar disk, sunspots follow trajectories that most of the time are curved and inclined; and their curvature and inclination exhibit precise seasonal changes that are repeated with an annual cycle.

On the other hand, Galileo did not pursue systematically the observation and study of sunspots after the publication of his book in 1613. There, he held that sunspots follow apparently rectilinear paths parallel to the plane of the ecliptic, and that the sun's rotational axis is perpendicular to that plane. However, in the *Dialogue*, he claims that within a year thereafter he happened to observe the curved path of a large sunspot, and that this observation led him to make a prediction, which he later verified by observation. His prediction was that the paths of sunspots are curved and inclined most of the time, and follow an annual cycle of changes. This prediction was based on the conjecture that the sun's rotational axis is inclined to the ecliptic and the Copernican assumption that the earth revolves annually around the sun. Thus, Galileo admits that he was originally wrong about the direction of the sun's axis, and that he has changed his mind in that regard. But he also claims that he had anticipated the main discovery announced by Scheiner in *Rosa Ursina*.

These claims exacerbated the priority dispute, expanding it to involve an additional detail about sunspots. It is beyond the scope of this book to elaborate this aspect of the controversy, since the focus of this passage in the *Dialogue* is the probative connection between the paths of sunspots and the earth's motion. However, I would sketch the situation as follows. There is evidence that Galileo had some knowledge of the inclination and curvature of sunspot trajectories well before the *Rosa Ursina* (Drake 1970: 177–99, esp. 183–84). There were at least two works published by other authors in 1626 that discussed this phenomenon (Besomi-Helbing 720–34, esp. 723–24). There is evidence that Galileo had seen the *Rosa Ursina* in 1631, before printing of the *Dialogue* was completed (Galilei 1890–1909, vol. 14: 294–95, 322; vol. 16: 391). Furthermore, it is undeniable that this passage represents the first time Galileo discussed this phenomenon. Finally, it is clear that in this passage, while agreeing with Scheiner on the observational facts, Galileo criticizes his (geostatic) explanation and advances an alternative (geokinetic) one. So, it seems safe to assert that Galileo had some direct acquaintance with, and some second-hand knowledge of, this phenomenon before the *Rosa Ursina*; but that this book made him aware of the robustness of the phenomenon; and that this passage was influenced (both positively and negatively) by Scheiner's work.

Galileo summarizes the observable facts about the trajectories followed by sunspots by means of some diagrams, which have been adapted with slight modifications. The first, Figure 6.3 (Smith 1985: 584, figure 1; cf. DML 404), represents four views of the solar disk and sunspot paths at roughly three-month intervals. MN is a line perpendicular to the line of sight of a terrestrial observer and located in the plane of the ecliptic, which plane may be imagined to be perpendicular to the plane of this paper, and in which plane we may regard the observer to be located. Most of the time, the trajectories are both inclined and curved with respect to the plane of the ecliptic.

However, twice a year, at about six-month intervals, their curvature is absent and their inclination is maximum, so that sunspots appear to follow a slanted rectilinear path (views A and C). In the first of these views (A), the slant is upwards in the sense that when a spot first becomes visible at E, it then appears to move along line EG toward G; whereas in the third view (C), the slant is downwards in the sense that spots that first become visible at G appear to move toward E along line GE.

Moreover, the curvature is maximum at two other times of the year (views B and D), also separated by six-month intervals from each other, but interspersed at three-month intervals with the two other views (A and C). At one of these times of maximum curvature (B), the curvature is upward from the plane of the ecliptic, but six

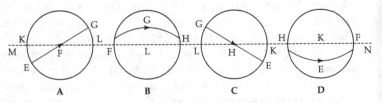

Figure 6.3

months later (D), the curvature is downward. Further, at these times the slant is absent in the sense that the beginning and ending points of a visible path (near F and H in view B; and near H and F in view D) appear equally distant from the ecliptic. At these times, the direction of motion is the same as for all other times, namely from the left to the right of the observer.

Aside from these four special times, the paths of sunspots usually appear both curved and slanted and are continuously changing, with the curvature being upward for half a year and downward for the other half, and with the slant also alternating in a similar manner.

Galileo does not say anything about the precise value of the inclination and curvature, nor about the precise times of the year when the various configurations appear. Such information had been given in *Rosa Ursina*. Thus this Galilean omission could be taken as an indication that his work was independent of Scheiner's. However, if Galileo were appropriating his rival's work, he would not have wanted to make his appropriation too obvious. The more important point to note is that Galileo's description focuses on the qualitative aspects of the phenomenon, and that for his purpose he does not need quantitative precision. His purpose is to argue in support of the earth's annual motion. In fact, the next step in his argument is to show that the variation and annual cycle of these trajectories could be explained as resulting from the earth's heliocentric revolution.

Galileo's geokinetic explanation of sunspot paths is best stated in terms of Figure 6.4 (Smith 1985: 545, figure 2). This diagram represents the earth's annual orbit (ABCD), or ecliptic, around the sun (KELG); but the earth's daily axial rotation is not represented, and plays no role in this part of the argument. AC and DB are diameters of the earth's orbit. NS is the axis of the sun's monthly rotation, and is inclined to the plane of the ecliptic; N and S are, respectively, the north and south poles. Circle EFGH is the solar equator, equidistant from the poles, and perpendicular to the axis. KFLH is a great circle of the solar sphere, at the intersection of the solar body and the ecliptic plane.

The key point is that as the earth revolves around the sun in a counterclockwise direction (through points A, B, C, D), terrestrial observers get different views of the path of sunspots. When (from D)

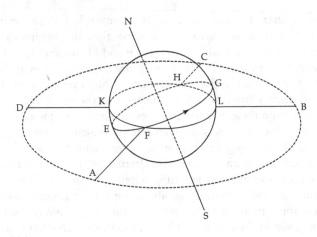

Figure 6.4

the sun's north pole is tilted toward us, sunspots rotating along the solar equator (or circles parallel to it) will appear to bend southward (i.e., downward), along curve HEF; whereas from B, when it's the sun's south pole that is tilted toward us, sunspots (still rotating counterclockwise over the solar body) will appear to bend northward (i.e., upward), along curve FGH. However, about a quarter of a circle after D, from A, we will see the greatest inclination northward (along EFG) but no bending; and from a point reached six months after A, namely from C, the path will again appear straight, still left-to-right or counterclockwise, but now inclining southward. All these visualizations become even easier if we compare this figure (embodying the geokinetic explanation) with Figure 6.3 (embodying the observational description), and if we note that the letters designating various geometrical points correspond.

After this explanation, Galileo goes on to admit that in a geostatic system one could also explain the changing annual pattern of sunspot paths. Clearly, this would involve attributing to the sun an additional motion. However, what this additional motion would be is not so clear.

In a geostatic, heliokinetic explanation, the most direct possibility would be to follow the standard Ptolemaic pattern that, for any motion observable in the heavens, there is a corresponding heavenly body undergoing that motion. This pattern contrasts with the Copernican one, according to which many apparent motions do not correspond to actual motions, but are merely the observational result of other processes; earlier we saw this pattern used for the case of retrograde planetary motion, and it also applies to sunspot paths. However, in the geostatic viewpoint, given that the paths appear to change, they are really changing.

To be specific, let us look again at Figure 6.3. In view A, the inclined axis of solar rotation has the north pole tilted to the left of the observer, and lies in a plane perpendicular to the line of sight. In view B, the north pole of the solar axis is tilted directly away from the observer, and lies in a plane along the line of sight and perpendicular to the ecliptic. In view C, the north pole is tilted directly to the right of the observer, and again lies in a plane perpendicular to the line of sight, i.e., in the opposite direction of the tilt in view A. And in view D, the northern axis is tilted toward the observer, and again lies in a plane along the line of sight and perpendicular to the ecliptic, i.e., in the opposite direction to the tilt in view B. That is, the north pole of the axis appears to be revolving clockwise along a circle which is located near the top of the solar disk, is parallel to the ecliptic, and is centered on a line perpendicular to the ecliptic and going through the center of the sun; and the south pole is revolving along a similar circle located near the bottom of the sun. From the point of view of the axis as a whole, the axis seems to be turning with a wobbling motion around the center of the sun in such a way as to describe two conical surfaces, one above and one below that center. Such a motion is called *precession*, and so we are saying that the solar rotational axis appears to be precessing clockwise and annually around another axis going through the center of the sun and perpendicular to the ecliptic. The geostatic explanation is that the observed changes in sunspot paths are the result of the sun's axis having such an annual precessional motion.

Now, let us recall that in the geostatic system the sun has three other motions: the daily westward (clockwise) motion, together will all the heavenly bodies, around the earth; the annual eastward (i.e., counterclockwise) revolution around the earth and along the ecliptic; and the monthly eastward (counterclockwise) rotation around its own inclined axis. Thus, the geostatic explanation of sunspot paths attributes to the sun a fourth motion. The most immediate question that arises is how this fourth motion is to be combined with the other three. The Aristotelians had an initially plausible answer. They utilized the same mechanism which Copernicus himself had used in attributing to the earth a "third" motion. As we saw earlier (IIC4), this was a precessional turning of the terrestrial axis, once a year, in a westward (clockwise) direction, allegedly needed to explain why this axis always remains parallel to itself. Copernicus's third motion may be represented in Figure 6.5.[3]

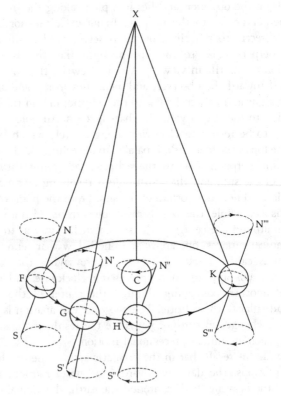

Figure 6.5

Here, C is the central sun in the Copernican system, and the earth revolves around it moving through points F, G, H, K, etc. Without the precession, as the earth moves in its orbit, its axis would constantly point toward point X on the celestial sphere, and so it would make a constant angle with the sun, as suggested by the four lines emanating from X and directed in turn toward F, G, H, and K. This would be a consequence of some rigid connection between the sun and the earth, i.e., of the earth being embedded in some solid crystalline sphere whose annual rotation around the sun carried the earth in a heliocentric revolution. However, with the axis precessing in the opposite direction to the annual motion, as the earth revolves around the sun, the axis wobbles just enough to maintain itself always parallel to its past or future self, as represented by the lines NS, N'S', N''S'', N'''S'''.

In the geostatic explanation of sunspot paths, one interchanges these Copernican positions of the sun and earth, placing the central earth at C, and the sun in an annual orbit thought points F, G, H, K, etc. If we make the solar axis of monthly rotation precess, this ensures that its north pole is pointing sometimes toward the earth, and six months later away from the earth, as well as sometimes to the right of the terrestrial observer, and six months later to the left. These configurations, in turn, would generate the observational features of sunspot paths.

This geostatic explanation of sunspot paths was basically the one that had been elaborated by Scheiner. Galileo does not explicitly attribute it to him, perhaps in order not to exacerbate the controversy; for his main purpose is to criticize it as being less preferable than the geokinetic explanation.

One disadvantage of the geostatic explanation can be seen from the point of view of the principle of simplicity: it postulates four motions as contrasted to the three of the geokinetic explanation. Moreover, it adds a second incongruity of motions: the first incongruity of the geostatic system was well known, namely that the direction of the annual motion is opposite to that of the diurnal motion; the second incongruity is a similar one specifically deriving from sunspots, namely that the sun rotates monthly on its axis in a counterclockwise direction, but its axis precesses annually in a clockwise direction. Neither of these two oppositions occurs in the geokinetic system: annual motion, diurnal motion, and solar monthly rotation are all in the same direction (counterclockwise, i.e., eastward), and there is no additional motion attributed to the sun.

A second reason for the superiority of the geokinetic explanation involves systemic coherence, or avoidance of *ad hoc*-ness. Attributing to the sun a fourth motion is an *ad hoc* maneuver, since it is postulated merely in order to explain the sunspot paths. By contrast, in the Copernican system, the annual motion, attributed to the earth for other reasons, is shown to be capable of explaining the new facts about sunspot paths.

A related, but distinct, advantage of the geokinetic explanation involves explanatory power. That is, the earth's heliocentric revolution can explain not only the changing shape and inclination of sunspot paths, but also the fact that the cycle of changes has a period of one year. This is simply a consequence of the fact that the period of the earth's revolution around the sun is one year. By contrast, in the geostatic system there is no explanation for this fact; in particular, it has no connection with the annual motion which it also attributes to the sun.

Finally, in Galileo's discussion there is also a hint that the geokinetic explanation has another advantage: greater predictive power. This is the methodological import of the story Galileo gives in this passage to claim priority of discovery over Scheiner. Aside from the controversial issue of historical accuracy, Galileo may be taken to be saying in part that the idea of the earth's motion led to his own discovery of sunspot paths, for it was this idea which enabled him to predict them, immediately after he had made an initial sporadic observation of a curved path.

In sum, Galileo claims that the geokinetic explanation of sunspot paths is better than the geostatic explanation because it has greater simplicity (fewer complications); more systemic coherence (less *ad hoc*-ness); greater explanatory power; and greater predictive power. Clearly he is not claiming that the geostatic explanation has to be excluded as impossible. He is explicit about two things: that the issue is one of satisfying "the discourse of the mind, as well as the course of the spots," and that the context is one of "conjectures and probable reasons."[4]

However, there is one final complication that needs to be clarified. Galileo is very cryptic about it, but his words contain all the hints for a reconstruction (DML 412). As previously mentioned (IIC4), Galileo argues in a later passage (IIIB4) that Copernicus's third motion is unnecessary. That is, as the earth revolves around the sun, the terrestrial axis of daily rotation can remain always parallel to itself simply by being left undisturbed; at least, this is so when we cut loose the rigid connection between the earth and the sun or the solid crystalline sphere carrying the earth around, and we regard these bodies (and all other heavenly bodies) as freely floating in some fluid medium. The constant parallelism of the terrestrial axis thus becomes an instance of rest.

Galileo is aware that such considerations can be applied to the sun's motion, if we are taking the geostatic viewpoint. Thus, the same considerations imply that the solar axis of monthly rotation does not need to precess in order to remain parallel to itself; it would automatically (spontaneously) do that. Thus, the sun's fourth motion is not really needed in a geostatic explanation of sunspot paths. We need not follow Scheiner in his adoption of Copernicus's own error. For as the sun revolves annually around the earth, and as the solar axis remains always parallel to itself, the result will be that at six-month intervals its north pole will alternate between pointing toward and away from the earth, and between pointing to the right and to the left of a terrestrial observer. This can be represented with Figure 6.6 (Smith 1985: 548, figure 6), where the four positions of the sun (A, B, C, D) are meant to correspond to the four observational views in Figure 6.3. The rest is self-explanatory.

Thus, one might think that the geostatic explanation of sunspot paths is comparable to the geokinetic explanation. For the former can do without the sun's fourth motion, and needs to postulate only three relevant motions, as is the case for the latter. Then, the geostatic explanation would lose its complexity, *ad hoc*-ness, and explanatory poverty vis-à-vis the geokinetic alternative. These considerations would not affect their relative status from the viewpoint of the fourth methodological principle mentioned above—predictive power. However, this would be a relatively weak advantage; moreover, its cogency depends on the accuracy of Galileo's story about his prior discovery of sunspot paths, which,

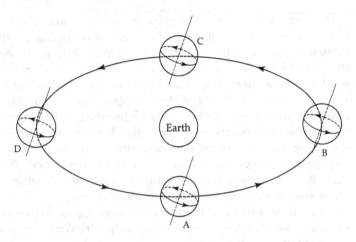

Figure 6.6

as mentioned earlier, is questionable, although perhaps not conclusively refutable. Unfortunately (for the geostatic explanation),
this is not the end of the (logical) story.

For in the geostatic system, the sun also has the diurnal motion
around the motionless earth. Now, if the sun revolves daily
around the earth, given that its axis of monthly rotation remains
parallel to itself, then in the period of 24 hours the orientation of
that axis relative to the earth would change continuously, dis-
playing in particular four special orientations defined as: north
solar pole pointing away from the earth, toward the earth, to the
right of the observer, and to the left of the observer. This would
mean that the continuously changing pattern of inclinations and
curvatures observed in the period of 12 months would have to be
observed every day. The main difference would be that the order
of the appearance would be reversed. Referring to Figure 6.3, the
sequence would be D, C, B, A, D, etc.; the reason for this is that
in a geostatic system, the diurnal motion is in a direction (clockwise, or westward) opposite that of the annual motion. However,
observation reveals no such daily variation is sunspot paths. Thus,
either the geostatic explanation stands refuted, or this lack of
daily variation has to be explained in some other way.

Now, it turns out that, in a geostatic system, there is a way of explaining the lack of daily variation in sunspot paths. To explain it, what must be devised is some means of ensuring that, as the sun moves in its diurnal orbit, during this 24-hour period the solar axis of monthly rotation always keeps its orientation toward the earth (not its parallelism to itself, which would be its natural, spontaneous tendency). This can be accomplished by making the solar axis precess on a daily basis, just enough to compensate the sun's diurnal rotation, and just enough to bring that axis back to being parallel to the position it displayed 24 hours earlier. And such a daily precession would have to be in a direction opposite to that of the diurnal motion; that is, it would have to be eastward, or counterclockwise.

Unfortunately, this would bring us back to attributing a fourth motion to the sun, and would render this version of the geostatic explanation at least as complex, *ad hoc*, and explanatorily poor as the previous geostatic version. Perhaps it would be even worse, insofar as one could now ask why the solar axis remains parallel to itself on an annual basis, but precesses on a daily basis. That is, in this version of the geostatic explanation, there would be one extra item which it would not be accounting for: not only would it have no explanation of why the period of precession for the solar axis should be 24 hours; but also, it would have no explanation of why the axis precesses on a daily basis, and not at all on an annual basis.

At this point, Scheiner or some other anti-Copernican could explain the lack of daily changes in sunspot paths by attributing the diurnal motion to the earth, rather than to the sun and all the other heavenly bodies. The solar axis would not have to precess daily, simply because the sun would not be revolving around the earth on a daily basis; thus the sun could keep its axis always parallel to itself in its annual orbital revolution. There would thus be no need of a fourth motion, for we would be postulating only a daily terrestrial rotation, a monthly solar rotation, and an annual solar revolution.

However, in attributing diurnal rotation to the earth, this position would no longer be geostatic, but partly geokinetic. It would represent a departure not only from Aristotle and Ptolemy, but also from Tycho, whose system was fully geostatic, although it made the planets revolve around the sun rather than the earth. Such a mixed explanation would perhaps gain something with regard to simplicity, number of motions, and incongruity of motions; but it would lose even more with regard to systemic coherence, since its *ad hoc*-ness would be more striking and more profound.

My conclusion is not that, as one recent scholar has claimed, "Galileo's proof is in fact absolutely conclusive" (Smith 1985: 544). Rather, I think that the geokinetic explanation of sunspot paths is absolutely better than any geostatic explanation, whether Ptolemaic or Tychonic, and whether the solar axis is rigidly connected to the earth or not. Galileo's explanation is also better than the mixed explanation that is both geokinetic and geocentric. Hence, Galileo's argument is very cogent, for it is an explanatory argument based on explaining sunspot paths, and showing that the geokinetic explanation is better than the alternatives.

IIIB2. DISTANCES AND SIZES OF STARS (DML 414–32)

In the last three sections, Galileo has presented three arguments in favor of the earth's annual revolution around the sun. Next, he resumes his critique of the arguments on the other side. He begins by mentioning the objection based on biblical passages (DML 415), on the pretext that Locher had included it in his book of *Mathematical Disquisitions*. However, Galileo barely mentions this argument, for it is obvious that he has no intention of discussing it in the *Dialogue*, since he knows very well that it is a forbidden topic, in light of the Index's anti-Copernican decree and the Inquisition's warning to him in 1616. Thus, he quickly drops that subject by saying that although "Holy Scripture ... is to be treated always with reverence and awe" (FIN 246; cf. DML 415), it is not proper to use it to attack someone engaged in hypothetical, natural, and human reasoning, as he is.

Returning to such reasoning, Galileo goes on to examine a group of astronomical objections that could not be refuted with new telescopic evidence, for the simple reason that the telescope did not at that time reveal the required phenomena.

These geostatic arguments were based on various claims about the fixed stars.

One particular argument involved the distance from the earth to the fixed stars and their size. The key point was that no annual changes in the appearance of fixed stars could be observed (with either the naked eye or the telescope); thus, if the earth had annual motion, then fixed stars would have to be extremely far away and extremely large; but such distances and sizes were regarded as absurd for various reasons; so the conclusion seemed inescapable that the earth could not have the annual motion.

The context of this argument was one in which there was general agreement about certain astronomical parameters, which can be approximated as follows: for the earth's radius (ER), 3,250 miles; for the sun's radius, 5 1/2 ER; for the radius of the annual orbit, namely the distance between the earth and the sun, 1,200 ER; and for the radius of the orbit of the outermost planet, Saturn, 9 times the radius of the annual orbit, or about 11,000 ER. Furthermore, regarding the fixed stars, proponents of the geostatic view accepted figures such as, approximately, the following: for the radius of a star of the first magnitude, about 5 ER; for the distance to a fixed star, about 20,000 ER, with some estimates as low as 14,000 ER (Van Helden 1985: 30, 50). However, Copernicans were committed to stellar distances and sizes about 1,000 times greater. Finally, note that according to present-day knowledge, the Copernican figures discussed at that time still represented gross underestimations of the correct amounts, since they relied on some traditional underestimated numbers and on relatively imprecise telescopic observations.

This disagreement about stellar dimensions was significant partly because stellar distances are a measure of the size of the universe. Moreover, disagreements about stellar distances raised the question whether there was any empty space between the outermost planet Saturn and the fixed stars; traditionally one heavenly region began where the previous smaller one ended, and so the inner surface of the stellar sphere was supposedly contiguous with the outer surface of Saturn's orb.

Galileo is aware that the Copernican stellar distances implied by the lack of annual changes in stellar appearances are very large—much greater than the traditional ones. In fact, in this section he clarifies how and why such stellar dimensions are very great, but also criticizes the claim that they are absurd. To see this, let's begin with a statement of the anti-Copernican argument from stellar dimensions:

> Let us suppose, as Copernicus himself says we must, that the earth's orbit, along which Copernicus makes it revolve around the sun in one year, were imperceptible compared to the immensity of the stellar sphere; then one would have to say necessarily that the fixed stars are at an unimaginable distance away from us, that the smaller ones among them are larger than the whole annual orbit, and that some others are much larger than the whole orbit of Saturn; but such dimensions are too large, as well as incomprehensible and incredible ... No one denies that the heavens can exceed in size anything we can imagine, or that God could have created the universe a thousand times greater than it is. However, we must admit that nothing has been created in vain and is useless in the universe. Now, when we see this beautiful order of the planets, arranged around the earth at appropriate distances to produce on it effects for our benefit, for what purpose should one then interpose between the farthest orbit of Saturn and the stellar sphere an extremely vast space which is without any star, superfluous, and in vain? For what purpose? For whose comfort and utility? ... To what end and for whose benefit do such large bodies exist? Are they perhaps produced for the earth, that is, for an extremely small point? And why are they so distant as to appear so small and as to be able to effect absolutely nothing on the earth?
>
> (FIN 248, 258–59, 262; cf. DML 416, 426, 430)

This formulation makes clear that the justification of the alleged absurdity of the Copernican stellar distances and sizes was a teleological and anthropocentric one, and that is precisely the aspect of the argument which Galileo criticizes. However, in this formulation it is not clear how one arrives at the claim that in the Copernican system the stellar dimensions had to have the very large values which were then declared to be absurd. There were different (but compatible) ways of arriving at such figures which were used by the authors Galileo has in mind: Locher, whose *Disquisitions*

(1614) is being explicitly targeted; Tycho, whom Galileo here mentions by name, and on whom Locher relied; and Ingoli, who is not explicitly mentioned here, but whom Galileo had criticized in his unpublished "Reply to Ingoli" (1624), along lines echoed in this passage.[5] Let us focus on Galileo's own computations, which, as already mentioned, may be regarded as a constructive clarification of their argument.

Galileo begins with the radical claim that all previous astronomers have grossly overestimated the visible apparent diameter of fixed stars by a factor of about 30. For they generally said that the apparent diameter of a first-magnitude star was 2 or 3 minutes of arc, whereas it is in fact no more than 5 seconds. And he thinks he can even account for this universal error. That is, under ordinary conditions, at night and with the naked eye, apparent stellar diameters are perceived much larger than they are primarily because of the irradiation effect on the eye, which makes us see them with a halo; this is the same disturbance which Galileo discussed earlier (IIIB1) for the case of the planets Venus and Mars. To reduce or eliminate this observational error, Galileo recommends observing a fixed star either with a telescope, or during the day, or with the help of a fine string behind which it can be made to hide. The effect of this correction is that, other things being equal, the actual size of stars in a Copernican universe can be reduced by a factor of 30. However, the Copernican numbers for stellar distances are still very large, as Galileo now goes on to compute.

His first computation involves a comparison of the apparent diameters of the sun and fixed stars. For this purpose, he utilizes the value of about 30 minutes (or 1/2 degree) for the apparent diameter of the sun; this was the commonly accepted figure, and was not affected by the irradiation effect because of the large size. Next he uses the value of 50 thirds (or 5/6 of a second of arc) for the apparent diameter of a fixed star of the sixth magnitude, which of course is smaller that that of a first-magnitude star. Then he calculates the ratio of the apparent diameter of the sun to that of such a star: this ratio is 2,160, since 30 minutes = 108,000 thirds (= 30 x 60 seconds per minute x 60 thirds per second), and 108,000:50 = 2,160. This means that assuming a star of the sixth magnitude is about the same actual size as the sun, such a star is 2,160 farther away (and that's why it appears 2,160 times smaller).

Galileo does not convert this figure to a number of earth radii, but let us do this conversion, since that is what the anti-Copernicans had in mind. Using the figure of 1,208 ER for the earth–sun distance, which he adopts from traditional astronomy, we get 2,609,280 ER as the distance to such a fixed star. This moves the discussion into the millions, which so much worried the anti-Copernicans. In fact, such a distance is more than one hundred times the traditional estimate of stellar distance, which as we saw was at most 20,000 ER.

However, as Galileo points out, such a million-size figure is still too small, for the following reason. He now considers how the appearance of such a star might change as a result of the earth's motion; here he does this in a preliminary or qualitative manner, along the lines his opponents were thinking of (although he will refine the analysis in the next section). They were thinking that the earth's annual motion would enable terrestrial observers to change their observational position by as much as the diameter of the earth's orbit; this diameter is, of course, twice the earth–sun distance. Given that we are talking about a star which is 2,160 times further away from the earth than the sun is, the earth's annual motion should produce changes in stellar appearance of the order of 2 parts in 2,160.

Now, it so happens that such a change is of the order of the change in the apparent position of the sun due to changes of position on the earth's surface. For, and as one moves on the earth from one location to a diametrically opposite one (or for a Copernican, as the daily spin takes an observer from sunrise to sunset), the observational location changes by 2 ER; but as we have seen, the sun is 1,208 ER from the earth; thus, we get a change in the appearance of the sun of 2 parts in 1,208. This is about twice the change in stellar appearance due to the earth's annual motion; but such a stellar parallax is about equal to the change in apparent solar position due to just one earth radius, namely 1 part in 1,208. And the important point about the latter is that it is observable; in fact, it is the so-called diurnal solar parallax (Dreyer 1953: 339). It follows that if fixed stars are 2,160 times farther than the sun, some stellar parallax should be observable. But it is not. Therefore, one had to either reject the earth's annual motion, or postulate even greater distances.

Galileo argues that such greater distances can be plausibly postulated. This can be done based on the correlation between size of orbit and period of revolution. He points out that many Ptolemaic astronomers held that there is some such correlation. This was well illustrated by the planetary revolutions, whose orbital radii increased as their periods increased. For example, using approximate round numbers, Saturn had an orbit with a radius of 9 times that of the annual orbit (AO) and a period of about 30 years; Jupiter an orbit with a radius of 5 AO and a period of 12 years; and Mars an orbital radius of 1.5 AO and a period of 2 years. Here Galileo does not try to formulate a single relationship that would cover all such instances, and which had in fact already been worked out by Kepler in his third law of planetary motion; but Galileo undoubtedly believed in some such law of revolution, for he held that Jupiter's satellites provided another independent and novel confirmation, as he suggested in another passage (IIA).

Thus, Galileo uses the parameters of these three planets to compute three separate estimates. In each case a comparison is made with the stellar sphere. Now, in the Ptolemaic system, because of the precession of the equinoxes, this sphere was attributed a rotation with a period of 36,000 years. The computation from Saturn's data is the following: the radius x of the stellar sphere is such that $x : 9$ AO $:: 36,000 : 30$ years; that is, $x = 10,800$ AO. Using Jupiter's parameters, we have a distance y such that $y : 5$ AO $:: 36,000 : 12$ years; that is, $y = 15,000$ AO. And with the Mars figures we get a distance z such that $z : 1.5$ AO $:: 36,000 : 2$ years; that is, $z = 27,000$ AO. These numbers are in terms of radii of the annual orbit, but in terms of earth radii the numbers would reach the millions (respectively, 13,046,400 ER; 18,120,000 ER; and 32,616,000 ER); and compared with the traditional geostatic estimates, these Copernican distances range from about 652 times greater to about 2,330 times greater.

The crucial point here is that, using the orbit-period correlations, the distance to the fixed stars is much greater than that computed from the (corrected) apparent diameters, ranging from about 5 times greater to about 12 times greater. At such distances, the annual change in the appearance of fixed stars would be correspondingly smaller than the diurnal solar parallax, and that would explain why it was not observed.

Such estimates are meant by Galileo to establish the plausibility, not the truth, of the large distances required in the Copernican world view; for such estimates are in the spirit of the Ptolemaic system, which has a stellar sphere rotating (precessing) around the motionless earth. However, such plausibility would remain even if one were to use a more correct value for the period of the precession of the equinoxes, which Copernicus estimated to be about 26,000 years; for in that case the computed stellar distance would be somewhat smaller (about 72%), but still large enough to imply an annual parallax much smaller (about 9 times) than the diurnal solar parallax. On the other hand, for Copernicus, the stellar sphere does not have a precessional rotation, since he explains the precession of the equinoxes as resulting from the precession of the terrestrial axis.

In light of such estimates, at this point an Aristotelian would resort to the teleological anthropocentric reasoning (quoted above), to show that such a great distance, and empty space, between Saturn and the fixed stars would be absurd and impossible. Galileo gives a memorable refutation that may be reconstructed as follows.

The concepts of *large* and *small* are relational, in the sense that to say that something is large or small makes sense only relative to a reference class. Hence, to say that the stellar sphere of the Copernican system is too large means that it is too large as compared with other heavenly spheres. But this is not so, because if we compare the Copernican stellar sphere to the sphere of the lunar orbit, the ratio is much smaller than that obtained from a comparison of an elephant to an ant, or of a whale to a gudgeon. Now, although the concept of size is subjective and relative in this way, it is not, however, teleological and anthropocentric. That is, it is wrong to think that the size of things fulfills a purpose definable in terms of human interests. To think so would be as absurd and arrogant as it would be for a grape to think, just because the sun acts on it in a way that is perfectly suited to cause its ripening, that the sun exists and acts only for its benefit. Similarly, it is wrong to argue that the universe cannot be as large as required in the Copernican system, because then the space between Saturn and the fixed stars would be superfluous and useless. Such an argument involves several non-sequiturs: to ground non-existence on superfluousness; superfluousness on lack of purpose for mankind; and lack of purpose for mankind on our ignorance of the purpose for mankind.

In sum, for Galileo, one cannot summarily dismiss the anti-Copernican argument based on the very great stellar sizes and distances implied by the failure to observe an annual stellar parallax. This argument is correct insofar as the stellar dimensions required in the Copernican system are indeed very great—even greater than those estimated by its opponents. And it is important to understand how and why such great distances are required. However, such great dimensions are not absurd because they can be shown to be comparable to familiar situations, both from common sense (such as the difference in size between the largest and the smallest animals) and from technical astronomy (such as the size of the stellar sphere implied by the orbit-period correlation). What is absurd is the attitude of teleological anthropocentrism, for it claims that everything that exists must have a purpose, definable by human interests, and known to human beings.

IIIB3. APPARENT POSITIONS AND MAGNITUDES OF STARS (DML 432–61)

Another version of the anti-Copernican argument from stellar appearances raised other issues, independent of the admissible distances and sizes of fixed stars. For the fact remained that no annual changes in the appearance of fixed stars could be observed, and everyone agreed that if the earth had an annual motion, there would have to be annual changes in stellar appearances; and from these two facts the conclusion seemed inescapable that the earth does not move annually around the sun.

However, when formulated in this manner, the argument is assuming that the non-observation of annual changes in stellar appearances implies that there are no such changes. Galileo thinks that this assumption is precisely the weak point of this argument. For although the non-observation of such changes may be due to their non-existence, "it may be that the immense distance of the stellar sphere renders such minute phenomena unobservable; it may also be ... that they have not even been investigated, or (if investigated) not investigated in an adequate manner, namely with the exactness necessary for such minute details; but this exactness is difficult to achieve, both for lack of proper astronomical instruments (which are subject to many variations) and for the faults of those who handle them with less diligence than necessary" (FIN 278; cf. DML 449). In fact, this section contains primarily a description of a research program with three main parts: a clarification of some common misunderstandings about what one is looking for; a theoretical elaboration of the complicated changes in stellar appearances that would result from the earth's annual motion; and some practical suggestions about observational techniques.

Galileo begins by clearing up some common misconceptions about the general kind of changes that would be produced by the earth's annual motion. One was that the elevation of the celestial pole above the horizon would change; this betrayed a failure to understand that if the earth revolved around the sun, the celestial pole would be defined by the terrestrial pole around which the earth would revolve, and the elevation of the latter can change only by moving around the earth's surface, and not by any motion of the whole earth (DML 433–36). Another confusion was that the annual change in the apparent position of fixed stars would be comparable to the large changes in stellar elevation above the horizon resulting by moving around the earth's surface; this misunderstood the difference between moving on a curved surface (the earth's) while measuring stellar elevations relative to that surface, and moving on a plane (the ecliptic) while measuring stellar elevations relative to that plane (DML 436–37). A third misunderstanding was that the annual changes for the fixed stars would be comparable to the large changes easily observable in the elevation of the sun over the horizon, which generate the cycle of the seasons; such thinking failed to appreciate the difference between moving in an orbit around a body—the earth around the sun—and moving in an orbit far away from a body—the earth vis-à-vis the fixed stars (DML 439–40, 452–61).

Next, we have the constructive theoretical elaboration of the annual changes in the appearance of the fixed stars that would result from the earth's annual revolution. Sagredo summarizes them clearly and succinctly:

> I think you have explained to us two kinds of different phenomena that could be observed in the fixed stars due to the earth's annual motion: the first is a variation in their apparent magnitude, depending on the extent to which we (carried by the earth) approach or recede from them; the other is their appearing to us sometimes at a higher and sometimes at a lower elevation along the same meridian (which also depends on the same approaching and receding). Moreover, you tell us (and I understand it very well) that these two kinds of variations do not take place equally in all the stars, but are larger in some, smaller in others, and nil in still others; the change in distance on account of which the same star should appear to us sometimes larger and sometimes smaller is imperceptible and almost null for the stars near the pole of the ecliptic, but is greatest for the stars lying in the plane of the ecliptic, and intermediate for those in between; the opposite happens in regard to the other variation, that is, the change in elevation is null for the stars lying in the plane of the ecliptic, greatest for those near the pole of the ecliptic, and intermediate for those in between. Furthermore, both of these variations are more noticeable in the stars which are closer, less noticeable in those which are farther, and would vanish in the extremely remote ones.
>
> (FIN 274–75; cf. DML 447)

To explain the details, let us refer to Figure 6.7.[6]

ANBO represents the earth's orbit in the plane of the ecliptic, which is imagined to be perpendicular to the plane of the paper and to be viewed at an angle by the reader. DFI and CEHJ are portions of two great circles which are perpendicular to the ecliptic plane, and along which are located some stars (C, D, E, F, H, I, and J) at various distances from the earth and at various

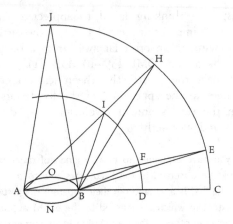

Figure 6.7

elevations above the ecliptic plane. AB is a diameter of the earth's orbit that has been extended to stars D and C.

First, consider stars (e.g., D, C) located in the ecliptic plane. At six-month intervals, the earth is located alternately at A and B. Looking at star C from A, a terrestrial observer will see it appearing smaller than from B, by an amount which is a function of the annual diameter AB and distance to the star, BC. The same will happen for star D. However, the change in the apparent size will be smaller for C than for D, because the change for C is a function of the ratio AB/AC, and the change for D is a function of the ratio AB/AD, and the former ratio is smaller than the latter. Moreover, there will be regular changes to the apparent angular separation of the two stars, as defined by the two lines from the earth to each of them: when the earth is at N, they are separated by angle CND; as the earth moves from N to B, this angle decreases to zero; as it moves from B to O the angle increases to its maximum value, COD; then as the earth moves from O to A the angle decreases again to zero; and as it moves from A to N the angle again increases to its maximum value.

Next, consider stars off the ecliptic plane, e.g., H. From A, star H will appear at angle HAB above the ecliptic; and from B, it appears at angle HBC. By visual inspection, one can see that the latter angle is greater than the former; that is, from B the star appears higher above the ecliptic than from A. By geometry, this difference has a definite value, which yields an important parameter. Considering triangle HAB, angle HBC is external, and hence equal to the sum of the two opposite angles, HAB and AHB; thus, HBC minus HAB equals AHB. That is, the annual change in the apparent position of a star, as defined by angular elevation above the ecliptic, is equal to the angle at the star, H, subtended by the diameter of the earth's orbit, AB. Calling that change *parallax*, we can say that the annual parallax of a star is equivalent to the angle which the diameter of the annual orbit subtends from the star.[7]

This yields another important result: annual stellar parallax is smaller for more distant stars and greater for nearer stars. Consider a star, I, nearer to the earth than H. The annual parallax of I is angle AIB. Once again, visual inspection reveals that it is larger than angle AHB, the annual parallax of star H. And geometry proves the same point: for considering triangle HIB, the external angle AIB is equal to the sum of the two opposite angles, IHB (= AHB) and IBH; thus, angle AIB is greater than AHB; i.e., the annual parallax of I is greater than that of H.

Now, contrast the positions of the three stars E, H, and J, and their annual parallaxes. By visual inspection, we can see that angle AEB is smaller than AHB, which in turn is smaller than AJB. There is also a geometrical proof of this, which Galileo gives (DML 445–46), but need not be reproduced here. But the general point is important: at constant distances, the annual parallax of stars increases or decreases as they are located at a greater or smaller elevation off the ecliptic. Moreover, this annual parallax reaches a maximum for stars, such as J, located at the pole of the ecliptic; and it vanishes to zero for stars, such as C, located on the ecliptic plane.

By contrast, the apparent size of a polar star such as J will not change in the course of the year, since its distance from the earth would not change; for the various distances (JA, JN, JB, JO) would be equal to one another. But as a star is located lower and lower off the ecliptic (H, E), there is some annual change in distance from the earth, and hence some annual change in apparent size; and these changes become greater and greater, reaching a maximum when the star, such as C, is located on the ecliptic plane.

Thus, the annual changes in stellar appearance are indeed numerous and complicated. Galileo claims that no astronomer has ever tried to elaborate them. Obviously, without an understanding of what changes to look for, the claim that no changes in stellar appearances have ever been observed loses its probative force. It does not justify the further claim that no such changes exist; and the anti-Copernican argument from stellar appearances is to that extent undermined.

Moreover, as it emerged in the last section, because of the great distances involved, such stellar changes would be extremely small, well beyond the power of then-available instruments. To reinforce this point, in this section Galileo discusses some evidence from a familiar situation of daily life. Suppose we are looking at a flaming torch at night from a distance of about 200 yards, and suppose the torch is moving at a relatively slow speed, being carried by a person walking. In the darkness of the night, we certainly could not tell whether the torch was moving toward or away from us; and for small distances, we could not discern even whether it was moving; for example, even after moving 20 yards, we could not tell the difference between the old and the new location. Then Galileo compares this situation to that of the astronomical observation of fixed stars, referring to some of the estimates discussed in the last section.

According to traditional views, the radius of Saturn's orbit is 9 times that of the annual orbit, i.e., the earth-sun distance; and a fixed star is about twice as far, i.e., 18 times farther from the earth than the sun. If we attribute the annual motion to the earth, the distance between the earth and a fixed star (located on the ecliptic) then changes by 2 parts in 18, or about 11%. Now, the undetectable change in distance of the flaming torch was 20:200 yards, or 10%. Thus, it seems reasonable that we cannot detect such a stellar change, just as we cannot detect such a change in the torch. Next, using the most conservative Galilean estimate of stellar distance calculated in the last section, namely 2,160 radii of the annual orbit, the earth's annual motion would change this by 2 parts in 2,160, namely, by about 0.0009, or about 0.1%. This change is about 100 times smaller than that of either the torch or a fixed star calculated from Ptolemaic estimates. Salviati is arguing that if we cannot notice a 10% change in the torch, it is not surprising that we cannot notice a Copernican change of 0.1% or less in the apparent diameter of a fixed star.

Such changes are not only very small, but also refer to quantities that are themselves very small. That is, they are changes of 0.1% or less in quantities of the order of 1 second of arc, which was Galileo's estimate of the apparent diameter of a star of the sixth magnitude. Thus, he senses (correctly) that to try to detect such changes would be hopeless.

However, he realizes that the prospects are more promising for the possible detection of the annual parallaxes, which are indeed extremely small changes, but refer to very large quantities (DML 448). For these are the angles of elevation off the ecliptic, which can reach 90 degrees; as one can see from Figure 6.7, angle AJB is small, but regardless of how small it is and how far star J is, angles CAJ and CBJ are close to 90 degrees. Thus, Galileo suggests that a good choice in the search for annual parallax would be Vega, which is a star close to the ecliptic pole.

This suggestion was indeed fruitful. For two centuries later, German astronomer Friedrich G.W. von Struve (1793–1864) made Vega the subject of extensive observations for the purpose of detecting the annual parallax, and he was able to detect it in 1839.[8] Thus, he became a co-discoverer of the stellar parallax, which had also been independently discovered a year earlier by another German astronomer, Friedrich W. Bessel (1784–1846). Bessel had focused on the star 61 Cygni using a different technique: measuring the annual change in the angular separation between this star and others that appeared nearby on the celestial sphere. And this was also envisaged by Galileo,[9] for it corresponds to his point that, for stars at different distances, the earth's annual motion would produce changes in their angular separation; he makes this point as applying to stars D and C, which are on the ecliptic plane, but it obviously applies also to other stars that are off the ecliptic, such as the pair F and E, and the pair I and H.

More generally, Galileo's suggestions in this passage were closely followed in the early eighteenth century by English astronomer James Bradley (1693–1762), although his observations led him to discover in 1729 a phenomenon called the aberration of starlight, instead of the sought-after parallax. He observed that the apparent position of fixed stars does vary annually: when observed at six-month intervals, their apparent position shifts by about 40 seconds of arc. But he understood that this shift cannot be a parallax partly because the change is too large and is the same for all stars, and partly for more technical reasons involving directions of the shift and times of the year; rather it results from the combination of the earth's annual motion and the finiteness of the speed of light.

Finally, Galileo makes some practical suggestions about how one might be able to detect the very small changes in the angles that define the apparent position of stars. He says that what is needed is instruments whose sides are miles long, so that differences of seconds in stellar elevation correspond to distances of the order of cubits along the instrument. This would contrast to even the very best previous astronomical instruments, such as those of Tycho, where the celestial quantities being measured correspond to differences of a hairsbreadth on the instrument. Galileo is talking about using topographic features on the earth's surface, such as mountains. The idea is to observe the changes in the way a particular star would be hidden by some beam on top of a building, at the top of a mountain, by carrying out observations from the valley below, at different times of the year.

In sum, in this section Galileo criticizes an important argument against the earth's annual motion based on the fact that no annual changes in stellar appearances were observed. The criticism is partly negative and destructive, insofar as he points out that this argument presupposes that the non-observation of annual stellar changes is best explained by their non-existence; but this assumption is false because such non-observation is better explained by the failure to know what exactly should be observed, to observe systematically enough, and to use adequate instruments for measuring extremely small quantities. However, the criticism is primarily positive and constructive, in the sense that Galileo clarifies misconceptions about what to look for, elaborates the complex details of what might be observable, and makes practical suggestions about how to carry out the observations; and all this amounts to a research program that turned out to be fruitful, as subsequent astronomers succeeded in observing stellar aberration and parallax, thus providing direct evidence proving the earth's motion.

IIIB4. MAGNETS AND MULTIPLE NATURAL MOTIONS (DML 461–81)

Near the end of the Second Day (IIC4), we saw Galileo criticize the anti-Copernican argument based on the metaphysical principle that simple bodies cannot have more than one natural motion. His criticism included two very brief points: one involving Chiaramonti's sharing Copernicus's error about the earth's "third motion" (DML 304–5), the other being the empirical refutation of this principle with evidence about the three natural motions of magnetized bodies (DML 305–6). In this section, Galileo expands on his two earlier cryptic criticisms. He also gives a clearer statement of the part of the anti-Copernican argument justifying the conflict between this principle and the various motions attributed to the earth by Copernicus. Moreover, Galileo gives a more forceful criticism of that part of the argument. Finally, besides revisiting the multiple-motions argument in these ways, he explores the relevance of William Gilbert's work on magnetism.

In fact, this section could be regarded primarily as a critical appreciation of Gilbert's work, and thus mostly as a digression from the main argument of the *Dialogue*. By way of appreciation, Gilbert is praised for the pioneering spirit with which he investigated a relatively novel topic, and for the seriousness of his experimental approach. Moreover, Galileo adopts Gilbert's thesis that the earth as a whole is a magnet, and elaborates an investigation of his own inspired by Gilbert; that is, an explanation of why lodestones fitted with an armature can attract much more weight than without the armature (sometimes 80 times more). However, Galileo also expresses a few reservations: methodologically, Gilbert should have shown more appreciation of mathematics; and substantively one of his claims is untenable, namely that magnets have an additional natural motion, insofar as when freed of external disturbances a spherical magnet would spontaneously rotate around its own axis.

Our focus, however, is the argument from multiple natural motions. So, let us examine that strand of the discussion. Simplicio begins with the following statement of the objection:

> Now, since the whole model of Copernicus seems, in my opinion, to be built upon infirm foundations in that it relies upon the mobility of the Earth, if this should happen to be disproved, there would be no need of further dispute. And, to disprove this, the axiom of Aristotle is, in my judgment, most sufficient: That of one simple body one sole simple motion can be natural. But here in this case, to the Earth, a simple body, there are assigned three, if not four, motions, and all very different from each other. For, besides the motion as a heavy body towards the centre, which cannot be denied it, there is assigned to it a circular motion in a great circle about the Sun in a year and a vertiginous rotation about its own centre in twenty-four hours. And, what in the next place is more exorbitant, and which, haply, for that reason you pass over in silence, there is ascribed to it another revolution about its own centre, contrary to the former of twenty-four hours, and which finishes its period in a year. In this my understanding apprehends a very great contradiction.
>
> (Santillana 407; cf. DML 461–620)

Galileo begins his criticism with an analysis of these various terrestrial motions to determine whether there is really a violation of the Aristotelian axiom. First, downward fall is a motion of the parts of the earth toward its center, not a motion of the whole terrestrial globe. The whole earth has never been observed to move toward its own center; and what's more, it is logically impossible that it should do so, because the motion of a whole moving toward its own center is conceptually incoherent. Here Galileo is repeating a point he had made in his critique of the geocentric argument from natural motion (IA4). Since the other motions mentioned by Simplicio do involve the whole earth, the Aristotelian axiom should not be applied to downward fall; otherwise we would have a situation in which the multiple natural motion would belong to different bodies, which is not excluded by the axiom.

Second, consider the daily axial rotation and the annual heliocentric revolution, which do belong to the whole earth. Perhaps these two processes do not contradict the metaphysical axiom because they could be regarded as a single motion. To understand this, consider an analogy. Consider a ball rolling down an inclined plane. In this process one can distinguish a spinning of the ball around itself and an advance of the whole ball along the surface of the plane. These two aspects of the process are certainly distinguishable, but perhaps they should not be separated; perhaps they should not be hypostatized into separate entities. Similarly, one can conceive a ball rolling counterclockwise around the outer rim of a wheel; such a ball would be rotating counterclockwise around itself, and simultaneously revolving counterclockwise around the center of the wheel. If something analogous is going on with terrestrial rotation and revolution, then these two things may not have enough ontological reality to make their existence conflict with the metaphysical axiom.

Next, we come to the third motion which Copernicus erroneously attributed to the earth, and which Galileo mentioned (without elaboration) in the critique of Chiaramonti's objection from multiple natural motions (IIC4, DML 304–5) and in the presentation of his own argument from sunspot paths (IIIC3, DML 412). Here Galileo discusses this topic more explicitly, clarifying why Copernicus's "third motion" of the earth does not exist but is an instance of rest (and hence that the principle of the impossibility of multiple natural motions cannot be properly applied to it). We explained this important clarification above, in our discussion of the paths of sunspots, and so we need not repeat it here. However, the present section contains some additional considerations, which are worth summarizing.

Consider the following experiment. Fill a bucket with water and place on its surface a floating ball. Go to the middle of a room and hold the bucket in your hands, while you move your feet in such a way as to turn your body around by 360 degrees in a counterclockwise direction. This makes the bucket, the water, and the ball revolve in a circle around your body. Be sure to observe carefully what happens to the ball. As the bucket moves counterclockwise around you, the ball will be seen to rotate in the opposite direction (clockwise), relative to the surface of the water. However, such a clockwise rotation by the ball is not real, but a kind of optical illusion. For what is really happening is that the floating ball is retaining its orientation relative to the walls of the room. Your moving the bucket and the water around yourself is really making the water turn around the ball. So the appearance of the ball's clockwise rotation is the result of the counterclockwise revolution of the bucket-water-ball system and the ball's tendency to aim itself toward the same point of the room's wall.

Apply these considerations to the earth's revolution around the sun, while it simultaneously rotates around its own axis in the same (counterclockwise) direction. Recall that the terrestrial axis of rotation would be inclined to the plane of the earth's orbit. Like the ball in the bucket experiment, this axis would always point in the same direction. But imagine yourself located at the sun and looking at the earth; and focus your attention on the earth's axis. As the earth revolved around the sun, it would seem to you that the north pole of the axis was undergoing an annual (clockwise) wobble, sometimes pointing toward you, then to your left, then away from you, then to your right, and then again toward you. But this phenomenon would be produced by the fact that the axis was *not* precessing relative to the celestial sphere. Thus, in the Copernican system, while the earth revolves yearly and rotates daily, its axis of rotation does not precess, yearly or daily; it remains always parallel to itself.

Galileo thinks that this feature of the earth's motion is a natural tendency of "every pendant and balanced body carried around in the circumference of a circle" (DML 462), as suggested by the bucket experiment. But he also gives another reason for thinking that the earth's motion would have this property: the whole earth is a magnet, and magnets have a tendency to point always in the same direction. This is how the topic of magnetism and Gilbert's work is brought into the discussion, then occasioning various digressions.

However, the properties of magnets provide Galileo with material for a more direct criticism of the argument from multiple natural motions; that is, more direct than the criticism elaborated so far in this section, which amounts to questioning the applicability of Aristotle's metaphysical axiom to the case of the earth's motion. As mentioned earlier, the more direct criticism is that this axiom is empirically false because magnetized bodies have at least three natural motions: downward fall, north-south alignment, and inclination from the horizontal plane. Gilbert's work adds substance to this claim about magnets.

There remains a clarification about whether magnets are simple bodies. For when faced with this direct empirical criticism, Simplicio objects that the Aristotelian principle refers to simple bodies. It is simple bodies that cannot have multiple natural motions, and such a formulation allows that mixed bodies may have multiple natural motions. Unfortunately, this line of reasoning contradicts the Aristotelian readiness to apply the principle to the earth and its motions. For such an application presupposes that the earth is a simple body, and this implies that magnets are similarly simple; for if magnets were mixed bodies, so would be the earth, since the whole earth contains magnetized bodies as parts.

In sum, Gilbert's work on magnetism enables Galileo to reinforce his criticism of the anti-Copernican argument from the impossibility of multiple natural motions. The multiple motions of magnets strengthen the empirical refutation of the Aristotelian metaphysical principle that simple bodies can have only one natural motion. Additionally, the thesis that the earth is a magnet strengthens the claim that, if the earth were in motion, its axis would always point in the same direction relative to the celestial sphere, thus eliminating the need for the earth's third (precessional) motion, and thus undermining the applicability of the metaphysical principle. However, Galileo also had some astronomical evidence for the empirical falsehood of the principle, and additional reasons for the inapplicability of the principle to the case of the earth's motion.

NOTES

1 Cf. *Sunspots* (in Galilei 1890–1909, vol. 5: 196–97; Galilei and Scheiner 2010: 262–63); *Assayer*, no. 18 (in Galilei 1890–1909, vol. 6: 273–76).

2 Cf. Besomi-Helbing 720–60, Drake 1970: 180–96, Heilbron 2010: 280–85, Hutchison 1990, Langford 1966: 124–25, Mueller 2000, Smith 1985, Topper (1999, 2000, 2003).

3 Adapted with some modification from Smith 1985: 546, figure 5.

4 Respectively, FAV 380, 383. Cf. respectively: DML 410, Santillana 362; and DML 413, Santillana 365.

5 Locher 1614: 25–27, Brahe (1596: 190, 1602: 481), Bucciantini and Camerota 2009: 171–73, Galileo's "Reply to Ingoli" (in Galilei 1890–1909, vol. 6: 523–33, translated in Finocchiaro 1989: 166–74); cf. Besomi-Helbing 762–80.

6 Adapted with modifications from FAV 412 and DML 447.

7 A more common definition of the annual stellar parallax equates it to one half of this, namely the angle at the star subtended by the radius (not diameter) of the earth's orbit.

8 DML 572–73 n., DCA 488 n., Besomi-Helbing 796.

9 Strauss 562 n. 76 (= Sexl-Meyenn 562 n. 76); Gapaillard 1993: 257, 269–70; Besomi-Helbing 792.

7

DAY IV

GEOKINETIC EXPLANATION OF TIDES

IVA. CLARIFICATION OF PROJECT AND CRITICISM OF ALTERNATIVES (DML 483–92, 535–39)

The Fourth Day examines a topic which Galileo has been hinting at on several previous occasions (DML 6, 244, 323, 479), i.e., the problem of explaining why the tides take place. He makes it clear at the outset that he is going to propose to explain this phenomenon in terms of the earth's motion, and that this explanation will provide a novel argument in favor of the Copernican hypothesis that the earth moves.

Galileo also makes it clear that this tidal argument for the earth's motion is different from his previous pro-Copernican arguments in the following important sense. It involves physical considerations about terrestrial phenomena, whereas those others have involved astronomical considerations about heavenly bodies: apparent diurnal motion (IIA), the negative correlation between luminosity and mobility (IIC5), the heliocentrism of planetary revolutions (IIIC1), retrograde planetary motions (IIIC2), and the paths of sunspots (IIIC3).

Recall also that terrestrial physical phenomena were extensively discussed in the Second Day, in the context of criticizing the mechanical objections to the earth's motion. However, that criticism showed only that the physical phenomena being appealed to could take place on a moving earth, not that they provide evidence in favor; it merely refuted the anti-Copernican arguments, without providing pro-Copernican ones. On the other hand, the tidal motion of the sea is different from the other mechanical phenomena: whereas they could occur on both a motionless and moving earth, it could only occur on a moving earth; and this is what yields a pro-Copernican argument.

Additionally, Galileo makes the methodological clarification that, from another viewpoint, the tidal argument has something in common with the earlier pro-Copernican arguments. Echoing his terminology, they all "confirm" the earth's motion by way of providing what are variously called indications, traces, clues, hints, intimations, attestations, and evidences.[1] Of course, some arguments are better than others, the top ones being those based on planetary heliocentrism, retrograde motions, sunspot paths, and tides; and among these stronger ones, the tidal argument is the strongest, in Galileo's opinion. Apparently, it is not a matter of proving the earth's motion, with an apodictic demonstration, like the proof of a mathematical truth, in which it is self-contradictory to assert the premises and deny the conclusion. It is important to keep this clarification in mind, since in the middle of the argument Galileo sometimes uses language that might convey the impression of a demonstrative proof.

This methodological clarification also corresponds to the message conveyed in the final passage of the Fourth Day (DML 537–39). That is, God is all-powerful and all-knowing, and so He could have created any number of possible worlds, different from the actual one; thus, regardless of how much good evidence there is in favor of some thesis about physical reality, it would be wrong to claim that this thesis must be necessarily true, for that would deny divine omnipotence; therefore, physical truths are always contingent, and can be denied without self-contradiction. We know that Galileo was required to include this argument, much beloved by Pope Urban VIII, in a prominent place in the book.

However, this external constraint does not undermine the logical cogency of this argument, or Galileo's ability to admit this cogency.

Besides these methodological clarifications, we have a descriptive review of the most basic facts about the tides and a critical review of the main alternative explanations. The phenomenon of the tides consists of three kinds of periodic motions of seawater: vertically up and down motion, such as the rising and falling of the water level visible inside a harbor, along the sides of a pier; horizontal back and forth motion, such as the water currents observable in certain straits; and a combination of these two, such as the ebb and flow that takes place on some beaches and shallow coastlines, where the water alternates between both rising and flowing inland, and both dropping and flowing outward to sea.

Moreover, these tidal motions occur in accordance with three cycles. The diurnal period, which is by far the most noticeable one, refers to the fact that usually the water rises or flows in one direction for about six hours, and then drops or flows in the opposite direction for another six hours, and so on; thus, during the twenty-four hours of a single day, there are two high and two low tides separated by six-hour intervals, and two full cycles of twelve hours during each of which the water exhibits all the motions that characterize a particular location. The monthly period refers to the fact that in the course of a month, on some days the high and low tides range farther than on other days. Finally, the annual period refers to the fact that in the course of a year there is a similar variation in high and low tides: during some months the high and low tides are greater than during other months.

This phenomenon puzzled mariners and thinkers from times immemorial, and various attempts were made to explain why seawater exhibits such motions. According to one theory, the tides are caused by differences in the depth of the water, the idea being that deeper water pushes and displaces more shallow water (cf. Besomi-Helbing 840–41, Aristotle 354a25–27). Galileo dismisses this explanation as misconceived, for such displacements obviously cannot be produced by differences in depth (which certainly exist), but only by differences in the level of the water surface (which do not exist).

Another explanation claimed that the tides originate from the existence of gigantic caves with trapped air at the bottom of the oceans:[2] as the weight of the water compresses such air, the level of seawater drops; as such compression overshoots the equilibrium point and becomes too great, the air starts to expand and makes the seawater rise; but again, the expansion proceeds past the equilibrium point for a while, until the process begins to repeat itself; the result is a perpetual oscillation that generates the tidal motions. Galileo's objection to this explanation is that such oscillation would produce the same tidal motion in all parts of a particular sea basin, e.g., both in Venice at the northwestern end of the Adriatic Sea and in Dubrovnik at its southeastern end; but that is not the case.

Next, some had tried to explain the tides as resulting from an attraction exerted by the moon toward the water directly below.[3] This explanation was partly supported by the well-known fact that there is a correlation between the daily motion of the moon and the tides. However, Galileo dismissed this explanation for two reasons. First (DML 487), he raised the same objection as for the previous explanation; that is, lunar attraction would produce the same tidal motion in all parts of a given sea. Second (DML 535–36), he regarded the lunar-attraction account as methodologically inappropriate, insofar as attraction was an "occult" property involving a magical view of nature and the resulting explanation as tantamount to explaining a phenomenon by giving a name to it. Thus, at one point, he even chides Kepler for his "childish" belief in a lunar-attraction theory. The irony here is that the lunar-attraction explanation turned out to be essentially correct, when Newton (1687) explained the tides in terms of the law of universal gravitation and the different gravitational forces exerted primarily by the moon (but also by the sun) on different parts of the oceans and land.

A variant of the lunar-attraction theory tried to take some of the occultness out of the lunar influence (cf. Besomi-Helbing 843). It claimed that the moon heated the water directly below it on earth; this made the water expand and rise in the way it does during a high tide. Then as the moon moved away from the meridian of a particular sea, the water cooled and contracted, and its level dropped. Galileo objects that this would imply that water during high tides is warmer and less dense than water during low tides, and this is not true.

Finally (DML 490–92), Galileo generalizes his criticism of these four particular explanations by claiming that the key flaw for all of them is the attempt to account for the tides while keeping the earth motionless; i.e., to produce tidal motions without moving the basins holding the seawater, to explain a dynamic phenomenon statically. From his particular critiques, he concludes that it is physically impossible for tides to occur on a motionless earth.

At this point, Simplicio interjects that if none of these theories can explain the tides, since the earth does not move and the geokinetic explanation is false, he would be inclined to say that the tides are produced by a divine miracle—a direct intervention of God into the natural order. Galileo has an ingenious criticism of this appeal to a miraculous intervention. He does not deny the possibility of miracles; in a sense, all divine intervention into the natural order is miraculous. The crucial question is, what miracle are we appealing to? One possible miracle is for God to intervene countless times for every single instance of tidal motion that occurs, and at the same time ensure that such tidal motions do not move the motionless earth. But another possible miracle is for God to intervene just once, by giving the earth the Copernican motions, and then let the tidal motions follow naturally. Which of these two miracles should we attribute to God? Galileo thinks the latter miracle is the one we should attribute to God, because it is simpler. In short, the miracle-explanation of tides is not incompatible with the geokinetic account; rather, the latter is a special case of the former. In this case, the task is to determine whether the tides could result from the earth's motion.

IVB. EXPLAINING THE DIURNAL PERIOD OF TIDES (DML 492–506)

The first step in Galileo's explanation is to show that when a container of water undergoes acceleration or retardation, the water experiences vertical up and down motions at the extremities and horizontal back and forth motion in the middle part of the container. Labeling such movements tidal-like motions, and taking the term acceleration in the strict general sense that includes retardation as negative acceleration, we may say that the first step is to support the *generalization that acceleration causes tidal-like motion.*

Consider a boat with a large rectangular tank full of water, going from one seaport to another over a calm sea, such as one of the boats that deliver fresh drinking water to the city of Venice from the nearby coast. Consider now what happens to the water in the tank as the boat begins its journey and starts moving. The acceleration imparted to the boat and the tank is not instantly communicated to the water in the tank. This water will tend to be left behind and flow backwards; so its level will rise at the back end of the tank and drop at the front end. Being a fluid, the water will then tend to flow in the opposite direction, from the back to the front end; the water level will then rise in front and drop in the back. This process of oscillation will continue for a while, even though the boat may have reached a uniform cruising speed. During this process, if we look at the middle part of the tank, the water there will not rise or drop, but rather move horizontally backwards first, then forward, then backwards again, and so on. After a while, however, if the boat continues moving uniformly over a calm sea, the water in the tank will also calm down and no longer experience those tidal-like motions; it will simply follow the forward motion of the boat.

Next, suppose the boat runs aground in shallow water, thus experiencing a strong reduction in speed until it stops. Such retardation will cause the water in the tank to start its oscillatory motion, like during the initial acceleration, except in reverse order. At first, the water will rush forward, increasing the level in front and lowering it at the back; then the water will move backwards, increasing the level at the rear and decreasing it at the front end. This change in water level at the extremities, together with the back and forth horizontal flow in the middle, will continue for a while. But eventually such oscillatory motion will again stop, and the water in the tank will calm down.

Besides such acceleration from rest to a certain constant cruising speed, and from the latter back to rest, merely increasing or decreasing the speed of the boat while cruising would have similar effects.

Galileo's first intermediate conclusion is that the acceleration or retardation of a body of water causes it to undergo tidal-like motions, such as vertical rising and falling of its level at the extremities of the container, and horizontal back and forth flow near the middle. Next, he applies this generalization to the case of a moving earth in the Copernican system. The key analogy here is that a sea basin, e.g., the Mediterranean Sea, is a container of water being carried by the moving earth. The generalization applies in the sense that the earth's axial rotation and orbital revolution combine in such a way as to generate daily accelerations and retardations in the motion of such a sea basin. The second step in the Galilean explanation is precisely to show that, if the earth were moving (with the two Copernican motions), every point on earth would regularly and alternately undergo a daily acceleration and retardation.

In Figure 7.1, circle GEC is the orbit in which the earth moves in a counterclockwise direction, around the central sun A. Circle DEFG represents the earth, which rotates around its own center B, in the same (counterclockwise) direction. Here both motions are regarded as constant: the earth's center B moves around the sun A at a speed which enables it to traverse the whole circumference of

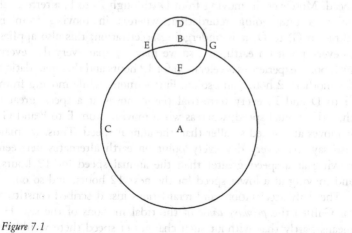

Figure 7.1

circle GEC in one year; while any given point on the earth, for example D, in the period of 24 hours traverses the whole terrestrial circumference, DEFG.

Now let us combine these two motions with each other. For a terrestrial point located on the opposite side from the sun and experiencing midnight, e.g., D, the diurnal speed and the annual speed are in the same direction, toward the left; thus they add up to give point D an actual speed that is the sum of the two, by reference to absolute space or at least to the center of the sun A. But for a terrestrial point located on the side facing the sun and experiencing noon, e.g., F, the annual and diurnal speeds are in opposite directions, the former still toward the left but the latter toward the right; thus they work against each other to give point F an actual speed that is the difference between the (greater) annual speed and the (smaller) diurnal speed. For intermediate points on the earth's surface (E, G), their diurnal speed has no effect on their annual speed, at least from the point of view of the west-to-east (i.e., right-to-left) direction; thus, such points move toward the left with a speed equal to just the annual speed.

Other features should be noted. At D the total actual speed is the maximum, at F the minimum; this applies to every point on the earth's surface in the course of 24 hours, and so any such point will alternate between such a maximum and minimum speed. Moreover, in moving from D (through E) to F, a terrestrial point is undergoing retardation; whereas in moving from F (through G) to D, it is undergoing acceleration; this also applies to every point on earth, and so we may say that every day every such point experiences acceleration for 12 hours and then retardation for another 12 hours, and so on. Furthermore, while moving from G to D and E, every terrestrial point moves at a speed greater than the annual speed; whereas when moving from E to F and G, it moves at a speed smaller than the annual speed. Thus, we may also say that every day every point on earth alternates between moving at a speed greater than the annual speed for 12 hours, and moving at a lower speed for the next 12 hours, and so on.

The daily accelerations and retardations just described constitute for Galileo the *primary cause* of the tidal motions of the sea. He means partly that without such changes of speed there would be no tides; i.e., they provide a necessary condition for the tides. Moreover, he also means that such changes of speed get the whole process of the tides started; in this sense, they contribute to the production of the tides. However, by itself the primary cause is not sufficient, and we need other concomitant causes, which he subdivides into secondary and tertiary. The secondary causes involve the fluid properties of water, which Galileo elaborates at length. The tertiary causes are such factors as the flow of large rivers into small seas, the action of winds, and the topographical interrelationships of sea and land.

The *fluid properties of water that act as secondary causes* are the following. First, water has a tendency to oscillate before reaching equilibrium. Second, the period of oscillation varies depending on the length of the sea basin, such that the period is shorter when the length is smaller. Third, the period also varies as a function of water depth, such that the period is shorter when the depth is greater. Fourth, as we have already seen, water in a given basin has the ability to move vertically at the extremities and horizontally in the middle. Finally, different parts of the same body of water can move at different speeds simultaneously.

An example of the last property may be seen from Figure 7.1. Suppose there is a very large sea, extending in length about a quarter of the earth's circumference, from D to G. In such a sea, the part near D would be undergoing great speed, the maximum, in the leftward direction; but the part near G would be experiencing an intermediate leftward speed. The difference would be even greater for a sea that extended over half the earth's circumference, say from D to G to F; for the part near D would be moving (leftward) at maximum speed, and the part near F at minimum speed, the difference being twice the diurnal speed.

After this general account of the secondary concomitant causes, Galileo proceeds to discuss a number of particular phenomena, which I call *secondary tidal effects*. These are tidal phenomena different from the basic tidal motions, and can all be seen to result from the interaction of the primary cause, the secondary concomitant causes, and the tertiary concomitant causes.

To begin with, there are no tides in lakes and small seas. Galileo says this is due to two factors: given the small size, the acceleration acquired from the earth's motion by such a body of water changes very little from one part to another, and so there is very little effect produced by the fluid property that different parts of the same body of water move simultaneously at different speeds; the other factor is the interference between the 12-hour interval of acceleration and retardation stemming from the primary cause and the much shorter period of oscillation deriving from the length of the water basin. Similarly, there are no significant tides in seas, like the Red Sea, which are narrow in an east-west direction but long in a north-south direction; the explanation is analogous to the previous one, but to see the analogy, we must recall that the acceleration due to the primary cause acts in an east-west direction, and so the size that counts for a sea basin is the size in an east-west direction.

Another secondary tidal effect is the strong currents that pass through certain straits, such as the Strait of Messina, separating the island of Sicily to the south from the Italian mainland to the north and connecting the Ionian sea to the east and the Tyrrhenian sea to the west; such currents derive from the horizontal back and forth motion produced by the primary cause in the middle of a sea basin, combined with the fluid property that to move greater quantities of water through a given channel greater speed is required. A related phenomenon is the violent agitations observable in certain straits, such as the Strait of Magellan, connecting the South Atlantic Ocean to the east and the South Pacific Ocean to the west and separating the mainland of South America to the north from the island of Tierra del Fuego to the south; here we see partly the action of the constriction factor mentioned in the previous explanation, but more importantly the mutual interference of the oscillation periods characterizing the two interconnected oceans.

An effect that illustrates the action of a tertiary cause is the unidirectional flow of currents in certain straits. One of these is the Bosporus, a narrow and long channel separating Europe to the west from Asia Minor to the east, and connecting the Black Sea to the north and the rest of the Mediterranean Sea to the south. In the Bosporus, water flows always southward. This happens because there are several large rivers flowing into the relatively small Black Sea, including the Volga, the largest and longest river in Europe; moreover, the primary cause is inoperative here, because the Bosporus is very narrow in an east-west direction.

This discussion of secondary tidal effects is interesting in itself, and revealing for what it tells us about the complexity of the tidal phenomenon and of the Galilean explanation. However, I have not yet mentioned what may be the most crucial detail, namely the six-hour interval between high and low tides, i.e., the occurrence of two high tides and two low tides each day. Galileo regards this detail as a secondary effect. That is, he admits that the primary cause alone would be insufficient to produce this detail, since the primary cause generates a maximum speed and a minimum speed only once a day, separated by twelve-hour intervals, during which there is continuous acceleration following the minimum speed, and continuous retardation following the maximum speed. Thus, he sketches the explanation of this detail by adding a concomitant secondary cause, namely the oscillation period of the water in the Mediterranean basin, which interferes with the primary cause to yield the observed six-hour interval. In accordance with such an appeal to a secondary cause, Galileo seems to think that the six-hour interval probably characterizes only the Mediterranean Sea as a whole, but that there are smaller seas within the Mediterranean that have a shorter period, and that longer periods (with about 12-hour intervals) occur in parts of the Atlantic Ocean.[4]

By treating the six-hour interval in this manner, Galileo is implicitly denying that it is a constituent element of the basic tidal motions and periods. These basic tidal effects are, as we have phrased them: vertical up and down motions at the extremities of a sea basin; horizontal back and forth motions in the middle; and a daily period, in the sense that the cycle lasts 24 hours before the phenomena start repeating themselves. He thinks that the combination of the two terrestrial motions enables us to understand how, in a general way, these basic phenomena are generated. Then deviations from, or refinements of, these basics (such as lack of tides in small seas, the six-hour interval, etc.) are accounted for by adding concomitant secondary and tertiary causes to the primary one.

Finally, this detail provides a good introduction to the question of how to evaluate this Galilean argument for the earth's motion based on the explanation of the tides. For the interval between high and low tides is generally about six hours, and even in Galileo's time some people were aware of this fact, so he could be criticized for failing to investigate this particular detail more carefully. If he had admitted the generality of the six-hour interval, he would have had to regard it as part of the basic tidal phenomenon, and hence the conflict between it and the period of the primary cause would have been more serious. Then he might have been led to revise or reject his basic theory. However, it would be incorrect to object, as some critics do, that his theory is empirically false because it flatly contradicts the observation of the six-hour interval; this objection would be wrong because, as we have seen, Galileo was aware of the tension between his primary cause and the six-hour interval, and he tried to deal with it.

Another evaluative issue is this. Some people say that we have known since Newton in 1687 that the primary cause of the tides is the gravitational attraction of the moon, and so the Galilean theory is erroneous. There is some irony here, since, as we saw earlier (IVA), Galileo did mention the lunar-attraction explanation, but dismissed it as unscientific. However, in pondering such a criticism, one must be careful to avoid the superficialities and exaggerations which such critics themselves commit. For example: the lunar-attraction theories formulated before and after Newtonian mechanics are very different; to say that Galileo's theory is partly erroneous is not the same as saying that it is completely erroneous; conversely, to say that Newton's theory is essentially correct does not mean that it is completely correct; and to claim that Galileo's explanation of the tides is false (insofar as it is false), does not tell us what is wrong with the reasoning in his supporting argument.

For example, some critics attribute to Galileo the following error of reasoning. The tides involve motions of seawater that are obviously relative to the earth, i.e., the frame of reference in which tidal motions take place is clearly a terrestrial one. However, the acceleration produced by the combination of the earth's axial rotation and orbital revolution is relative to absolute space, or, more concretely, relative to the center of the sun; that is, the frame of reference of such changes of speed is a solar one. Galileo is allegedly confusing these two frames and illegitimately switching from one to the other. In particular, the analogy between the moving earth and the water-carrying boat fails; for in the case of the boat there is a single frame of reference that defines both the accelerations of the boat and the sloshing of the water in the tank, namely the frame of reference of the land masses. Whether this criticism is accurate or not, at least it does address an issue of reasoning. However, the criticism is questionable because it seems to me that even for the case of the earth and the tides, a single frame of reference is involved, namely the solar one.

Such issues cannot be pursued any further here, let alone resolved. Nor can many other issues be even mentioned. Suffice it to have just introduced such a fascinating topic.[5]

IVC. AIR MOVEMENTS AND TRADE WINDS (DML 506–16)

We have seen that Galileo's tidal argument for the earth's motion is multi-faceted: the first step involves the generalization that acceleration causes tidal-like motions; the second elaborates the primary cause, which lies in the acceleration produced daily by the combination of the earth's rotation and revolution; the third is a description of the secondary concomitant causes, which involve the fluid properties of water; and the fourth is an explanation of various secondary tidal effects, as due to the interaction of the primary, secondary, and tertiary causes. However, this is not all, for the argument also needs to be defended from objections. He does not regard the six-hour interval between high and low tides as an objection, but rather as a secondary effect which his account can explain. On the other hand, there are two objections which he does consider; both involve the earth's atmosphere.

The first objection is an internal criticism, advanced by Simplicio. That is, like water, air is a fluid and does not have to follow in every way all the motions which a moving earth would try to impart to it; thus, if the earth's motion causes tidal-like motions in the watery part of the terrestrial globe, then it would also cause similar motions in the earth's atmosphere. For example, in the case of air, one effect would be a constant wind from the east. Since no such wind exists, it follows that the earth's motion is not the cause of the tides and does not exist.

Galileo replies that there are two things wrong with this objection. One is that although there are some similarities between water and air, there is one crucial difference: water can conserve the motion it acquires relatively easily, whereas air cannot do it as easily. This property of water is instrumental in the mechanism whereby the earth's motion makes seawater accelerate, and this acceleration in turn produces the tides. Hence, the geokinetic mechanism that produces the tides would not have to produce similar motions in the earth's atmosphere.

This reply refutes one of the key premises of Simplicio's objection. Additionally, Galileo has a reply to the other premise regarding the lack of a constant wind from the east. He points out that this claim is false: as a matter of fact, there are prevailing winds from the east, especially in the equatorial regions. These are the so-called trade winds, which sailors know about and exploit in the art of navigation.

Besides contributing to undermining Simplicio's internal criticism of the tidal argument, the existence of the trade winds enables Galileo to formulate a new argument in favor of the earth's motion. It is this. On a rotating earth, the air surrounding the earth's surface would be mostly carried along, but would be left behind to some extent, because of the property that air does not conserve acquired motion very easily. In particular, the lower atmosphere would have more of a tendency to be carried along, partly insofar as it is surrounded by mountains, and partly insofar as it would have mixed with it some water (such as water vapor and clouds) and some of the element earth (such as dust). However, parts of the earth are characterized by large expanses of flat, non-mountainous surface that would not carry along the air as easily. The equatorial regions have this feature; additionally, their diurnal rotational speed is greatest. Thus, in such regions, there would be a greater tendency for the air to be left behind, which would generate a wind in a direction opposite to the terrestrial eastward rotation, namely a westward wind (or wind from the east).

Galileo makes no big issue out of this argument from trade winds, whose formulation occurs in the context of the evaluation of the tidal argument; so it is easy to miss it. However, Galileo is explicit, although modestly realistic, that with the evidence of the trade winds, "one confirms the rotation of the earth with a new argument taken from the air" (FAV 464; cf. DML 509), and "one can see how the phenomena of the air seem to correspond wonderfully with those of the water and with the observations of the heavens to confirm the mobility of our terrestrial globe" (FAV 465–66; cf. DML 511).

Although this argument (like all the others in the *Dialogue*) should be evaluated primarily in light of the historical context and the knowledge available then, here it is useful to add an anachronistic note. From the point of view of present-day knowledge, Galileo was right in claiming that the existence of the trade winds confirms terrestrial rotation. However, the connection between these two phenomena is not the semi-Aristotelian one he sketches in this argument.

Rather, the process gets started as the heat in the equatorial regions makes the air rise, thus causing cooler air from higher latitudes to rush toward the equator to take the place of that warmer air. Then, such cooler air moves toward the equator, tendentially in a direct north-south direction. But its eastward linear speed (stemming from the earth's rotation) is conserved, and so as it enters regions nearer the equator that have a higher eastward speed, it lags behind, i.e., it is deflected westward. In the northern hemisphere, the combination of the original southward motion and this westward deflection yields a motion toward the southwest, i.e., a wind blowing from the northeast. In the southern hemisphere the original motion (toward the equator) is *northward*, but the deflection (opposite to terrestrial rotation) is still westward, thus generating winds blowing from the southeast. In modern physics, this phenomenon is the coriolis effect, which I have mentioned before, in the discussion of north-south gunshots (IIB8) and of bodies falling from the moon (IIC1). In those two discussions, Galileo seems to have had an inkling of the coriolis effect, but that is not the case in the present passage.

Now, moving on to a more contextual evaluation of the argument from trade winds, Galileo does this himself by discussing an objection. That is, Simplicio tries to give an alternative explanation of the phenomenon, and his explanation is not only geostatic, but also tries to exploit the trade winds as an intermediate cause of the tides. This objection is therefore also a second criticism of the tidal argument.

Given that there are prevailing winds from the east (the trade winds—more or less), Simplicio argues that this could be explained in the Ptolemaic world view by the diurnal rotation of the lunar orb. The mechanism would be that as the lunar orb rotates daily around the motionless earth, it transfers some of this motion to the sphere of the element fire, which is located just below the lunar orb and just above the sphere of the air; then as the sphere of fire moves westward to some extent, it transfers some of its motion to the air. Admittedly, this westward air flow is mostly not transmitted to the lowest part of the atmosphere, which is surrounded by mountains. However, in the equatorial regions of the oceans, this lower atmosphere would acquire some of the westward flow, for the same reasons mentioned earlier in the geokinetic explanation of the trade winds. The trade winds would then be generated, thus providing an alternative to the geokinetic explanation, and hence a criticism of the wind argument for the earth's motion.

But that is not all. For such prevailing winds would, in turn, transfer some of their motion to the ocean water, which would thus acquire some westward motion. However, the land masses of the various continents would interfere with this water flow, partly by obstructing it. Moreover, this interference would partly force the westward flowing water to reverse direction, in a manner analogous to the downhill flow of rivers; there, the topographical features of the river banks and bottom sometimes cause parts of the river water to flow backwards, and other parts to form whirlpools. Here then, we have also explained the occurrence of the tides, thus providing an alternative to the geokinetic explanation, and a second criticism of the tidal argument. And such geostatic explanations of the trade winds and tides are more economical and less *ad hoc* than the geokinetic explanations, because the former do not introduce any novel process, as the latter does with the earth's motion.

Galileo's assessment of this objection is nuanced and memorable enough to be worth quoting:

One cannot deny that your argument is ingenious and possesses considerable probability; but I say probability in appearance, not in actuality and reality. It has two parts: in the first, it gives the reason for the continuous motion of the breeze from the east, and also for a similar motion of the water; in the second, from the same source it wants to provide the cause of the tides. The first part has (as I said) some semblance of probability, but very much less than what we get from terrestrial motion; the second is not only wholly improbable, but absolutely false and impossible.

(FAV 468; cf. DML 514)

Against the first part, Galileo argues as follows. First, it is doubtful that there really exists a lunar orb, made of the invisible crystalline aether, to which the lunar globe is attached, and whose rotation carries the moon in its geocentric motions; and recently even many Aristotelians have rejected such heavenly orbs, for all the planets, as well as for the moon; and they have accepted that the heavens are fluid, in the sense of being mostly empty, or filled with some substance of extremely low density, so as to allow the heavenly bodies to move through it. Second, it is even more doubtful that there really is an element fire, collected just under the lunar orb and above the air. Third, even if they both existed, the lunar orb could not transmit any motion to the sphere of fire, nor could the latter transfer it to the air; for, in each case, the two surfaces in contact with each other are supposedly very smooth, and additionally the motion would have to be transferred from the less to the more dense.

To refute the second part, Galileo says the following. Assume, for the sake of the argument, that the ocean water could acquire some westward motion, through transfer from the lunar orb to the sphere of fire and from the latter to the air. However, the westward moving ocean water, together with the interference of the continental land masses, could not produce tidal-like motions. For a key property of the tides is that at the same terrestrial location water moves regularly in opposite directions at different times, vertically up and down, and horizontally back and forth; whereas, as the evidence from river flow shows, the obstruction by the land to the water tending to move constantly westward could indeed cause water to reverse direction in some locations, but the reversal would be constant for any particular location. Thus, there would be no way of generating the key property of tides.

In sum, the properties of the earth's atmosphere cannot be used to refute the tidal argument. Such a refutation is attempted with the internal criticism that if the earth's motion causes tidal motions in seawater, it would have to cause similar tidal-like motions in the air. There is also the external criticism that one can give an alternative explanation of the tides, in terms of the diurnal motion of the lunar orb being transmitted first to the sphere of fire, then to air, and then to seawater. Both the internal criticism and the alternative explanation are untenable. On the contrary, the properties of the air can be used to formulate a novel argument confirming the earth's motion, based on the existence of the trade winds, and explaining their occurrence in terms of the earth's motion.

IVD. MONTHLY AND ANNUAL PERIODS OF TIDES (DML 516–35)

As already mentioned (IVA), tides have a monthly period: every month there are days when the high and low tides are greater than average, and other days when they are smaller than average. Similarly, tides undergo an annual period: each year there are times when the range of high and low tides is greater than normal, and other times when it is less. The last substantive discussion in the *Dialogue* provides explanations of these periods.

Galileo approaches this problem in terms of the principle of concomitant variations of cause and effect. That is, suppose that the existence of some effect has been explained as due primarily to some cause. If it also happens that the effect varies in a regular manner, it is best to explain this variation in terms of the same primary cause, by means of corresponding variations in it, rather than in terms of some additional novel cause. The reason is that such an explanation would be simpler, more economical, and less *ad hoc*.

Applying this principle to the tides, Galileo recalls that he has explained the basic tidal phenomenon as due primarily to the daily acceleration generated by the combination of the earth's axial rotation and orbital revolution. Thus, he proposes to search for variations in the terrestrial rotation or revolution (or both) that would produce the monthly and annual variations in the amplitude of the daily tidal motions. He takes up the monthly period first.

Galileo thinks he has found a regular monthly variation in the earth's annual motion. This would be a consequence of the general law that whenever a constant force causes a body to move in a circular path, the motion is slower when the circle is larger, and faster when the circle is smaller. This principle was discussed by Galileo on two previous occasions, where I labeled it the law of revolution: in one of the simplicity arguments for terrestrial rotation (IIA), and in estimating the distances and sizes of stars (IIIB2). Now he reiterates the supporting evidence from the orbits and periods of the planets and Jupiter's satellites, and adds other evidence.

Some of this additional evidence comes from the functioning of wheel clocks, whose timekeeping is adjusted by sliding a small weight along the diameter of a wheel. The more important additional evidence involves the behavior of pendulums, which Galileo had studied extensively, discovering some physical laws. He exploits the present occasion to devote several pages to this new branch of science. Here we need not follow the details, or even summarize his discussion, except for the most relevant law: the frequency of oscillation of a pendulum depends inversely on the length of the string, and on nothing else (such as weight and amplitude); thus, pendulums oscillate faster when the hanging weight moves along circles of smaller radius.

Next, Galileo applies this law to the joint motion of the earth and moon around the sun. As the earth revolves annually around the sun, the moon keeps up with the earth by revolving monthly around it. It's as if the earth and the moon form a system of two bodies that revolve around the sun as a unit. Now, as was well known from the uncontroversial explanation of the lunar phases, in a month the configuration of earth, moon, and sun changes such that the moon is sometimes in opposition to the sun (at full moon) and sometimes in conjunction with it (at new moon). This change in configuration is then viewed as a change of the effective radius of the circular path along which the two-body system of earth and moon moves around the sun: the effective heliocentric orbit is larger at full moon and smaller at new moon. Thus, by the law of revolution, near full moon the speed along the annual orbit is slower, and near new moon it is faster. In short, because of the changing positions of the moon, the earth's annual motion is not uniform, but undergoes monthly variations, being faster at new moon, and slower at full moon.

This variation in the earth's annual speed in turn generates variations in the acceleration produced by the combination of the annual and diurnal motions. For let us now recall the basic Galilean mechanism of the primary cause, in a situation where the diurnal speed is constant, but the annual speed changes: when the constant value of the diurnal speed is added to or subtracted from a higher annual speed (at new moon), the combination yields smaller relative changes in total speed, i.e., smaller accelerations; conversely, when the constant diurnal speed is added to or subtracted from a lower annual speed (at full moon), the combination yields larger relative changes in total speed, or larger accelerations. Correspondingly, the tides would be larger near full moon, and smaller near new moon.

After this exposition, Galileo defends this account from an objection. If the earth underwent this monthly variation in its orbital speed, the phenomenon would have been observed by astronomers, since the change would be reflected in the apparent speed of the sun in its motion along the zodiac; but this solar change has not been observed; therefore, there is no such variation in the earth's annual motion. Galileo replies by sketching a capsule history of astronomy, suggesting that many phenomena remain unobserved at any one time; there are various reasons for this: the investigator's failure to search for the phenomenon, the very small magnitude of the phenomenon, etc. Such is the case for the non-uniformity of the annual motion, and so its reality should not be doubted because it has not been observed so far; instead, one should devise means to test for it.

Finally, Galileo gives the explanation of why the magnitude of the tides follows an annual cycle of changes; it lies in certain seasonal changes involving the earth's diurnal motion. This is not to say that there is an actual variation in the speed of terrestrial rotation, but rather that there is an important consequence of the fact that the axis of rotation is inclined to the plane of the earth's orbit (by a constant amount of 23.5 degrees). So far, this inclination has not been taken into account, but when it is, the constant diurnal rotation has different effects at different times of the year, depending on the orientation of the earth's axis relative to the sun. The key difference is between those times or orientations called solstices and those called equinoxes.

In Figure 7.2, the earth is represented as revolving around the sun in a counterclockwise direction, while it simultaneously rotates daily in the same direction, such that the inclined axis of rotation remains always parallel to itself. The summer solstice (SS) is the time of the year when, and the location of the orbit where, the north pole is pointed toward the sun by the greatest amount, equal to the 23.5 degrees of the axis's inclination to the ecliptic plane. The winter solstice (WS) occurs six months later and half an orbit away, when the north pole points away from the sun by the greatest amount. The vernal equinox (VE) and the autumnal equinox (AE) occur at times and locations that are halfway between the solstices, when the earth's axis points neither toward nor away from the sun, but rather to the right and to the left, respectively, as viewed by an observer at the sun.

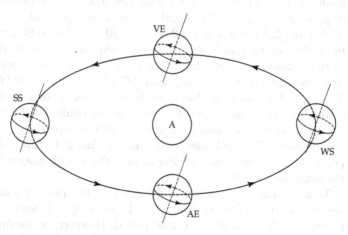

Figure 7.2

Given these terminological definitions and spatial configurations, let us refer back to Figure 7.1 (IVB) that illustrated the basic Galilean mechanism of the primary cause. The earlier discussion proceeded as if the circle of maximum diurnal speed, the equator DEFG, were in the same plane as the circle of the earth's heliocentric orbit, GEC. However, the plane of the equatorial circle is inclined by 23.5 degrees to the plane of the earth's orbit. When this inclination is taken into account, we can envisage the following modifications to that mechanism and figure, assuming we are viewing the situation from above the ecliptic plane, i.e., from the north celestial pole. At the summer solstice, the EFG half of the equator would lie below the plane GEC, whereas the GDE half would lie above that plane. At the winter solstice, the equator would be in a reverse configuration, with the EFG half lying above, and the GDE half lying below, the ecliptic plane GEC. Nevertheless, at both of these times, the component of the diurnal speed acting in the plane of the annual orbit is essentially identical to what it would be if the earth's axis were not inclined but perpendicular to the ecliptic.

For example, point D would be moving in the same direction as the annual motion (leftward) and parallel to it, and so the full amount of D's diurnal speed would be added to the annual speed; similarly, but in reverse, point F would be moving in a direction (rightward) opposite to the annual motion, although parallel to it, so that the full amount of F's diurnal speed would be subtracted from the annual speed. An equivalent way of looking at the situation would be to consider these motions and their combination from the viewpoint of the diameter EG of the equator DEFG. That is, at both solstices, during nighttime hours, a point on the earth's surface would move leftward and in tandem with the annual motion by an amount equivalent to the entire length of the diameter GE; and conversely, during daylight hours, the point would move rightward and against the annual motion by the same distance, EG.

Thus, at the solstices, the mechanism of the primary cause would function essentially unchanged, as compared with the previous explanation of the diurnal period. However, at the time of the equinoxes, there would be a difference, amounting to an effective reduction of the diurnal speed. To see how, let us consider once again Figure 7.1.

At the vernal equinox, with the north pole tilted to the right, the equator DEFG would also be tilted, in such a way that the DEF half would be above the ecliptic, and the FGD half below it. Thus, the diameter EG would also be inclined to the ecliptic (by 23.5 degrees). Furthermore, in the course of one night, while the diurnal rotation takes a point from G to D and to E, the amount by which the point has moved leftward is not equivalent to the full length of the diameter GE, but rather to its projection onto the ecliptic plane. The same would happen at the autumnal equinox, when the north pole is tilted to the left (as viewed from the sun), so that the DEF half of the equator would be below the ecliptic plane, and the FGD half above. Something similar would also happen during daylight hours, at both the vernal and the autumnal equinoxes: a point on the earth's surface would move rightward and against the annual motion by a reduced amount, defined as the component of the diameter along the ecliptic plane.

Galileo states that this reduction of effective diurnal speed at the equinoxes would be about 1/12. This would make the effective diurnal speed at the equinoxes about 92% of the actual diurnal speed. This corresponds to my own calculation, which found the projection onto a plane of a line inclined to it by 23.5 degrees to be equal to the length of the original line times the cosine of 23.5 degrees (which is 0.92).

Finally, the effect on tidal motions is this. At the equinoxes, the combination of annual revolution and diurnal rotation would consist of adding or subtracting a smaller diurnal speed from the same annual speed. This would yield smaller relative changes in speed, and hence smaller accelerations, and hence smaller tides. Hence, there would result an annual cycle such that tides would be smaller at the equinoxes than at the solstices and other times.

These explanations of the monthly and annual periods are certainly ingenious. Moreover, they possess considerable systemic coherence, stemming from the principle of concomitant variations of cause and effect. Furthermore, they have the merit of being mechanical, i.e., of utilizing reproducible processes of bodies in motion causing other bodies to move in various ways; this is especially true of the account of the monthly period, which treats the motions of the earth and moon like a pendulum or mechanical clock in the heavens. Such mechanical explanations may be contrasted to those described in terms of occult qualities, such as the power of a body to act at a distance by attracting another body.

Unfortunately, from the viewpoint of empirical accuracy, these explanations leave much to be desired. That is, they do account for the monthly and annual periods when these are described in a relatively vague manner, by saying that the magnitude of the tides is greater during certain days of a month and smaller during other days, and also that it is greater during some months of the year. However, they do not account for, and indeed contradict, the phenomena when these are described in a more precise manner. For observation of the monthly cycle reveals that the tides are larger at both full and new moon, and smaller at first and last quarter, whereas in Galileo's explanation they are larger only near full moon, and smaller near new moon. Similarly, observation of the annual cycle reveals that the tides are larger at the equinoxes, instead of being smaller as claimed in the Galilean explanation.

Galileo does not even discuss these discrepancies. Perhaps, he was not sufficiently well acquainted with the relevant observational details, although some of his contemporaries were. Or perhaps he was acquainted with them, but thought that these discrepancies could be accounted for by means of secondary and tertiary concomitant causes; this would be analogous to the way his explanation of the diurnal period tries to account for the six-hour interval between high and low tides. Instead, after presenting his explanation, he advances the following reservations, which may be an implicit clue to the difficulty. With these reservations, he ends his account of the tides:

> This is as much as I am able to tell you on this matter, and I dare say it is as much as it is possible for us to know with any certainty, since as you know, certain knowledge can [be] only [of] conclusions which are fixed and constant. Such are the three general periods of the tides, deriving as they do from causes which are one, invariable, and eternal. But these primary and universal causes are intermingled with

secondary [particular] causes which are capable of producing many changes. Some of these, such as the changing winds, are inconstant and unpredictable; others, such as the lengths of the sea's inlets, their different geographical orientations, and the great variations in the depth of water, are fixed and established, but [have not been] observed because they are so many and diverse. It would take very prolonged observations and absolutely reliable reports to compile an account of them which could provide a basis for firm suppositions about how they combine to produce all the appearances, not to say peculiarities and anomalies, which can be found in the motion of the tides. So I shall content myself with pointing out that such accidental effects occur in nature and are capable of producing extensive changes; I shall leave it to those who have practical knowledge of the various seas to observe them in detail ... such details require prolonged observations, which I have not so far [undertaken], nor do I expect to do so in future.

(Shea-Davie 354–55; cf. DML 533–35)[6]

NOTES

1 *Indizio, vestigio, confermazione, argomento,* and *attestazione* in FAV 442.17, 443.3–4, 462.24, 464.8, 464.28, 466.2, 466.33, 467.24, 487.9; *indications, trace, sign, confirmation, argument, footprints,* and *evidences* in DML 483, 484, 507, 509, 511, 512, 536 (= DCA 416, 417, 436, 439, 441, 462); *clues, hints, intimations, reason, confirmation, testimony, argument, track,* and *attestations* in Santillana 425, 443, 445, 448, 469; in French, *indices, trace, confirmation, argument,* and *témoignages,* in Fréreux-Gandt 405, 422, 424, 426, 442; and in Spanish, *indicios, vestigio, confirmación, argumento, pasos,* and *pruevas,* in Beltrán 359, 377, 378, 382, 401.

2 Cf. Strauss 567 (= Sexl-Meyenn 567), Sosio lxxiv, Besomi-Helbing 849–50.

3 Cf. Strauss 566–67 n. 3 (= Sexl-Meyenn 566–67 n. 3), Sosio lxxxi-lxxxii n. 2, Besomi-Helbing 841–42.

4 Cf. Galileo's "Discourse on Tides" (in Galilei 1890–1909, vol. 5: 388–89, and in Finocchiaro 1989: 128).

5 Cf. Aiton (1954, 1963, 1965), Besomi-Helbing 831–903, Brown 1976, Burstyn (1962, 1963, 1965), Clutton-Brock and Topper 2011, DiCanzio 1996: 339–52, Drake (1970: 200–213, 1979, 1983, 1986), Finocchiaro (1980: 6–24, 74–79; 2010b: 37–64, 235–43), Galileo's "Discourse on the Tides" (in Finocchiaro 1989: 119–33), Mach 1960: 263–64, McMullin 1967: 31–42, Naylor 2007, Palmieri 1998, Shea 1972: 172–89.

6 Here, to improve accuracy, I have emended the Shea-Davie translation in several places, which are indicated by my bracketed phrases.

Part III

SPECIAL ASPECTS OF THE *DIALOGUE*

8

SCIENCE
ROBUST CONFIRMATION OF EARTH'S MOTION

Several general but distinct aspects of the *Dialogue* remain to be examined. One is the book's scientific content and significance; this is already implicit in the earlier reconstruction of its main argument, but needs to be made explicit. Another is the book's methodological content and significance; this too has been partially included in that reconstruction, but merits a more comprehensive analysis. Finally, there is the topic of the book's rhetorical content, form and significance; this has barely been mentioned so far, but represents an important dimension. These three aspects will be examined, respectively, in the three chapters (8–10) of Part III.

8.1 SCIENTISTS' JUDGMENTS

Scientists from Newton to Albert Einstein (1879–1955) have found Galileo's *Dialogue* rich in scientific content and significance. For example, in the *Mathematical Principles of Natural Philosophy* (1687), Newton attributes to Galileo such fundamentals of modern science as the laws of inertia, force and acceleration, and times-squared free fall, and the principle of the composition (and resolution) of motion. He must have read these into the *Dialogue,* because he had already read this book but not Galileo's *Two New Sciences,* which might have been another possible source.[1]

In a different context, Einstein, too, had the occasion to express his own appreciation, when he wrote the Foreword to Stillman Drake's English translation of the *Dialogue* (1953). Einstein claims that the geostatic thesis was based on the hypothesis of the existence of an abstract center of the universe, and that this hypothesis was accepted because it provided an explanation of the fall of heavy bodies. Then he claims that Galileo rejected this hypothesis on the grounds that "although it accounts for the spherical shape of the earth it does not explain the spherical shape of the other heavenly bodies" (Einstein 1953: xiii). Einstein claims next that there is an analogy between this Galilean argument and the rejection, by his own theory of general relativity, of "the hypothesis of an inertial system for the explanation of the inertial behavior of matter" (Einstein 1953: xiii). The analogy is that both the Galilean and relativistic criticism object to the postulation of an entity with the very peculiar property of affecting the behavior of real objects without being affected by them, thus lacking the same kind of reality that matter or fields do. Finally, he asserts that such entities are "repugnant to the scientific instinct" (Einstein 1953: xiii).

Typical of scientists' attitude is the judgment expressed by Arthur Schuster (1916: 11) in his presidential address to the British Association for the Advancement of Science in 1915: "Modern science began, not at the date of this or that discovery, but on the day when Galileo decided to publish his *Dialogues.*"

Such readings by scientists are not always completely accurate from a textual, historical point of view. However, even when somewhat inaccurate, such scientists' judgments are important because they are essential in defining the book's scientific content and establishing the book's connection to the world of science. A good example of such an interpretation is Newton's attribution to Galileo of the second law of motion, that acceleration is proportional to the force causing it. In the *Dialogue,* this law is not found stated as clearly or used as explicitly as the laws of inertia and of squares, or the principle of composition. However, in the critique of the extrusion objection (IIB11), we do find the following statement: "two bodies of equal weight resist equally being set in motion at equal speeds; but, if one is to be moved faster than the other, it will offer a greater resistance corresponding to the greater speed it is to be given" (FIN 208–9; cf. DML 250). And it is not too far-fetched to see this statement as a groping toward Newton's law of force and acceleration (cf. Pagnini vol. 2: 413 n.).

In any case, *usually* scientists' readings *are* accurate, especially if we regard them as approximations, which are subject to various levels of correctness and refinement. Such is the case with the other three things which Newton found in the *Dialogue*.

Moreover, the scientific content which working scientists find in a book like the *Dialogue* does not always consist of specific truths, or factual statements, that correspond at some degree of approximation to the scientific knowledge of the current period. Even more relevant are ideas or theories, problems or questions, arguments, and methods or procedures found in the book that have some appropriate correspondence to those of current science. A good example of this is Einstein's reading, which involves an idea and a method of argument (see the discussion in Chapter 11.4 below). And this is the sort of thing Schuster had primarily in mind in his all-encompassing statement.

8.2 THREE-FOLD SCIENTIFIC ACHIEVEMENT

In one of his first references to the *Dialogue*, some twenty years before he published the book, Galileo described it as "an immense project, full of philosophy, astronomy, and geometry" (Galileo to Vinta, 7 May 1610, in Galilei 1890–1909, vol. 10: 351). Here, "geometry" means not only the science of lines and figures, but also the science of numbers; that is, he was referring broadly to what we would nowadays call mathematics. By philosophy he meant partly natural philosophy, the most basic branch of which was physics—the science of motion; he also meant what he would have called logic, which included not only the theory of reasoning, but also the theory of knowledge, method, and science. These philosophical theories correspond to what we would nowadays call logic, epistemology, methodology, and the philosophy of science; but for brevity's sake, here I will normally use one of these terms to refer interchangeably to any one or all four of these branches of philosophy, and so I will speak simply of epistemology, or methodology, or philosophy. The main distinction is between these and natural philosophy, for which I will normally use the term physics. In short, usually the word philosophy will refer not to natural philosophy, but rather to logic, epistemology, methodology, or philosophy of science.

With these terminological clarifications in mind, we can say that as early as 1610 Galileo thought of his *Dialogue* as a synthesis of astronomy, physics, mathematics, and epistemology. The correctness of such a description is relatively obvious from reading the actual book, and it is even more evident from my reconstruction above (Chapters 4–7). However, it will be useful to elaborate this aspect of the book more explicitly in this chapter, for doing so will enable us to appreciate the book's scientific content and significance. Moreover, focusing on such an interdisciplinary synthesis enables us to understand and appreciate another important part of the book's scientific achievement, namely the confirmation of the geokinetic theory.

This second, related aspect of the work was also clear to Galileo himself. In fact, in 1629, as he was about to finish writing the book, he wrote to a friend to convey the good news, and described it as "a very broad confirmation of the Copernican system" (Galileo to Diodati, 29 October 1629, in Galilei 1890–1909, vol. 14: 49). Here, the notion of confirmation is a very appropriate one; for it conveys the idea of strengthening a claim by rendering it firmer or stronger than it was before, or than some alternative claim, or more probable than not; and so it also conveys the idea of degrees of strength, firmness, probability, likelihood, or acceptability. This contrasts with the notion of demonstration or proof, which conveys the idea of conclusiveness, finality, or absoluteness.

Moreover, in speaking of the Copernican system, Galileo is referring primarily to the key claim that the earth moves, with daily axial rotation and annual heliocentric revolution. This is also obvious from the progression of the book's main argument (and of my reconstruction). That is, he does *not* mean to treat the Copernican system as a unit that can only be accepted or rejected as a whole. In fact, recall that Galileo rejects some parts of the system of Copernicus: for example, the "third motion" which Copernicus himself attributed to the earth (IIC4, IIIC3, IIIB4), and Copernicus's traditional conception of the celestial sphere as a real sphere on which all fixed stars are located at the same distance from the planetary system (IIIB3).

After the book was published, this geokinetic accomplishment was perceived not only by Galileo's followers, but also by critical peers. One of these was Descartes, who in 1634, after borrowing a copy of the book, which had already become a collector's item, wrote his impressions to a friend. Despite various reservations, on one key issue Descartes judged that Galileo "philosophizes pretty well about motion ... his reasons for the earth's motion are very good."[2]

There is a third way of describing the book's main scientific accomplishment. This corresponds to the point I made earlier (Chapter 1), when I said that critical reasoning represents not only the book's approach, but also its content. It also corresponds to the ecclesiastic restrictions under which Galileo was operating when he finally wrote the book. In short, the book is also a critical examination of the arguments on both sides of the Copernican controversy.

Recall that the main ecclesiastical restriction, as understood by Galileo, stipulated that he was not supposed to hold or defend the Copernican theory of the earth's motion. He must have felt that the book did not "hold" the geokinetic theory because it was not claiming that the geokinetic arguments were conclusive; and that the book was not "defending" the Copernican theory because it was a critical examination of the arguments on both sides. Recall also that there was a positive aspect to that main ecclesiastical restriction, namely that the earth's motion could be regarded as an hypothesis. Galileo must have felt that the book was an hypothetical discussion because the earth's motion was being presented as an hypothesis postulated to explain observed phenomena.

To be sure, the critical examination revealed that the geokinetic arguments were much stronger than the geostatic ones, and when the book's central thesis is so formulated, it is clear that Galileo holds and defends it, indeed that he demonstrates it successfully. However, this is a comparative, relative, and contextual thesis; the geokinetic opinion is discussed vis-à-vis the geostatic opinion and vis-à-vis the available evidence. Galileo must have felt that this discussion and this comparative thesis did not amount to holding or defending the geokinetic opinion in an absolute, objectionable, or illegitimate sense; at most it amounted to defending it as probable. Or at least, this was his gamble in writing and publishing the *Dialogue*.

In other words, the *Dialogue* discusses the earth's motion by examining all the arguments on both sides of the controversy. But the examination includes not only a presentation and an analysis of the arguments, but also an *evaluation* or *assessment* of their worth or strength. Galileo was indeed taking the liberty of evaluating the evidence; but he was hoping that if he carried out the evaluation fairly and validly, his having engaged in such assessment would not be held against him. Again, he was taking the gamble that a correct assessment of the arguments would not be seen as an objectionable defense of Copernicanism.

Similarly, it is clear that the book was not hypothetical in the sense of treating the earth's motion merely as a useful instrument of mathematical calculation and observational prediction. However, the book was clearly hypothetical in the sense that the earth's motion was being presented as an hypothesis useful for the understanding and explanation of observed phenomena. Thus, Galileo was taking the gamble that it would be enough for him to treat the earth's motion as an hypothesis in some significant sense.

Of course, we know that Galileo lost this gamble, for complicated reasons that involve the 1633 Inquisition proceedings. But the important point here is that such a "gamble" was quite consistent with his original project for a system of the world; indeed it was a brilliant actualization of what was feasible under the circumstances.

The *Dialogue* is, then, a synthesis of astronomy, physics, mathematics, and epistemology; a strong confirmation of the geokinetic hypothesis, as distinct from a demonstrative proof; and a presentation, analysis, and evaluation of all astronomical, physical, and epistemological arguments and evidence on both sides of the controversy. And these three aspects of the book are intimately related. The purpose of this chapter is to elaborate the substance of the synthesis, of the confirmation, of the critical examination, and of their connection.

8.3 CRITICAL EXAMINATION AND INTERDISCIPLINARY SYNTHESIS

Let us review the book's argument reconstructed earlier (Chapters 4–7) from the point of view of such an interdisciplinary synthesis, strong confirmation, and critical examination. The synthesis involves the four disciplines of physics, astronomy, philosophy, and mathematics. The critical examination involves two main aspects: the presentation and negative evaluation of the geostatic arguments and evidence; and the presentation and positive evaluation of the geokinetic arguments and evidence. And all this yields a strong confirmation insofar as all the relevant arguments and evidence are being considered, and insofar as the evaluations are fair and cogent.

Let us begin with the criticism of the physical arguments against the earth's motion. The argument based on the Aristotelian conception of natural motion is stated and criticized in sections IA2, IA3, IA4, and IIB1. Then there is the argument based on the principle that simple bodies can have only one natural motion; this is stated and criticized in sections IIC4 and IIIB4. The various arguments based on the behavior of falling bodies are formulated and criticized in sections IIB3, IIB4, IIB5, and IIC1. The several arguments based on the behavior of projectiles are critically examined in section IIB6, IIB7, IIB8, and IIB9. The objection from the flight of birds is refuted in section IIB10. And the argument from the extruding power of whirling is refuted in section IIB11.

In the course of this criticism, Galileo introduces and develops many of the principles of the new physics which he had discovered in the twenty years of his university career (1589–1609). In the critique of the Aristotelian argument from natural motion, while developing a more coherent conception of natural motion (IA3), Galileo elaborates some of his laws about motion on inclined planes; for example, the principle that two bodies descending along inclined planes of different inclinations acquire the same speed when the height of the inclined planes is the same (DML 25).[3] In the critique of the objections from falling bodies, from projectile motion, and from the extruding power of whirling (IIB4-IIB11), we find a principle of conservation of motion (which is an approximation to the law of inertia), and the principle of the composition, superposition, or resolution of motions. In the passage on falling to earth from the moon (IIC1), we have the laws of free fall, such as the law that the distance fallen increases as the square of the time elapsed, and the (equivalent) law that the velocity acquired by a freely-falling body is directly proportional to the time elapsed.

Next, let us consider the criticism of the astronomical objections to the earth's motion. The general argument from the earth–heaven dichotomy is criticized throughout the First Day, as well as in section IIIA. The criticism amounts to showing that this dichotomy is not only unjustified by the Aristotelian arguments (IA2, IA3, IA5, IA6, and IA7), but also empirically false (IA6, IB1, IB2, IB3, and IIIA). Although this criticism does yield the positive thesis that terrestrial and heavenly bodies are similar, it does not yield by itself a positive argument for the earth's motion. However, a related geokinetic argument is formulated in the discussion of the negative correlation between luminosity and mobility (IIC5).

The specific astronomical counter-evidence is criticized in the Third Day: the behavior and appearance of the moon and the planets Venus and Mars in section IIIB1; the distances, sizes, and apparent position, magnitude, and brightness of fixed stars in sections IIIB2 and IIIB3. In the last mentioned section, the objection from annual stellar parallax is not really refuted: although Galileo clarifies the problem considerably and outlines a fruitful research program, the lack of an observed parallax remains as a piece of counter-evidence. This is the only such case in the whole book, which justifies Galileo's cautious talk of confirmation. The lack of a perceptible annual stellar parallax diminished the degree of confirmation of the earth's motion, and hence of the substance of the book's scientific achievement; but in another sense there resulted an enhancement of its scientific content, insofar as that research program was taken up by many astronomers and eventually led to the successful detection of an annual stellar parallax, as well as to other results (such as the aberration of starlight).

In this criticism of the astronomical objections, Galileo introduces and develops the telescopic discoveries he had made during the period of his direct pursuit of the Copernican research program (1609–16): mountains on the moon; stellar composition of the Milky Way and the nebulas; existence of countless new fixed stars not visible with the naked eye; satellites of Jupiter; phases of Venus; regular variations in the apparent size and diameters of the planets; and sunspots, monthly axial rotation of the sun, and (perhaps) annual cycle of sunspot paths.

Let us now examine the philosophical objections, philosophical not in the sense of natural philosophy, but in the sense of epistemology, logic, methodology, and so on. The crucial objection from the deception of the senses is refuted in section IIC3. Then there are other, lesser arguments that are also examined: the teleological, anthropocentric argument in sections IA7 and IIIB2; and the objection from the alleged inexplicability of terrestrial rotation in section IIC2. Furthermore, in the course of my reconstruction, I pointed out (IIC4) that the argument from multiple natural motions has a metaphysical strand in it, besides the physical one. Finally, as mentioned earlier (Chapter 3.2), the objection from divine omnipotence has a logical aspect that is essentially correct; thus, Galileo's response to it (IVA) was not refutation or rejection, but rather clarification and utilization for a better self-reflective understanding.

With regard to mathematics, obviously the mathematics we are dealing with is applied, rather than pure, mathematics. There were not, nor could there have been, any purely mathematical arguments for or against the earth's motion. However, many arguments involved quantitative or geometrical considerations.

For example, in the criticism of the observational argument for heavenly unchangeability (IA6), one of Galileo's points is that the recent observation of sunspots shows that heavenly bodies are changeable. To prove his point, he has to show that these spots are really part of the body of the sun, rather than swarms of previously undetected planets circling the sun and obscuring our line of sight. And the crucial considerations in this proof involve the geometry of perspective: that when a body lying on the surface of a rotating sphere is viewed from a great distance, its apparent size and speed seem larger near the middle and smaller near the edges of the sphere's apparent disk. Similar, but much more complicated, multi-faceted, and intense considerations are found in the discussion of the apparent positions and magnitudes of the fixed stars (IIIB3), to derive exactly what changes would result from the earth's annual motion.

A common criticism advanced by Galileo against some anti-Copernican arguments is that they are quantitatively invalid. For example, in criticizing the argument from point-blank gunshots (IIB9), he shows that if we compute the amount by which westward shots should hit low, and eastward shots high, it is so small as to be undetectable. On a much more elaborate scale, his criticism of the argument from the extruding power of whirling (IIB11) points out that we need to compare and contrast the directions and amounts of extrusion and downward fall, and he argues that the extruding tendency along the tangent would never exceed the downward tendency along the secant. In criticizing Locher's argument from the time required to fall from the moon to the earth (IIC1), one objection made by Galileo is that Locher seems to presuppose the mathematical absurdity that the radius of a circle is longer that the circumference; and another criticism is to make a computation to arrive at a more accurate estimate of the time of fall. And in the discussion of the distances and sizes of fixed stars (IIIB2), Galileo makes several quantitative considerations aimed to compare and contrast various sizes, distances, and principles for estimating them.

In the discussion of the evidence about the 1572 nova (IIIA), Galileo's critical analysis of the available observational data is in a class by itself. Its key concern is to compare and contrast the numerical observational data of different astronomers; mathematically derive distances that are mutually contradictory; and resolve these contradictions by correcting the observational data with various quantitative rules aimed to minimize the total amount of corrections. One scholar has called it "a prescient application of probability, the first theory of error" (Swerdlow 1998: 266).

Let us now consider the favorable arguments. The most elaborate such argument is the one based on the explanation of the physical phenomenon of the tides in the Fourth Day, but clearly it is not the only one. The others are arguments based respectively on the following considerations: the annual cycle of sunspot paths (IIIC3); retrograde planetary motion (IIIC2); the heliocentrism of planetary revolutions (IIIC1); the simplicity of the earth's daily axial rotation vis-à-vis the daily geocentric revolution by the whole universe, for which he gave eight reasons (IIA); the negative correlation between the state of motion and the property of inherent luminosity of the heavenly bodies (IIC5); and the phenomenon of trade winds (IVC).

Here, first, it should be noted that the evidence mentioned in these geokinetic arguments is mostly astronomical, but not entirely, since the arguments from tides and from trade winds are appealing to physical terrestrial evidence. Accordingly, it is not surprising that in the course of elaborating the argument from tides, Galileo finds the occasion to introduce some of his discoveries in physics. In the explanation of the diurnal period of the tides (IVB), while elaborating some of the properties of fluid motion, Galileo formulates an approximation to an important law in hydrodynamics: that the period of oscillation of water in a vessel varies directly with the length of the vessel and inversely with its depth.[4] In the explanation of the monthly and annual periods (IVD), we find an elaboration of the laws of the motion of pendulums; for example, that the frequency of oscillation of a pendulum depends inversely on the length of the string, but does not depend on the weight of the oscillating body or on the amplitude of oscillation.

Second, except for two arguments, all the others have the logical structure of inferences to the best explanation; that is, arguments in which the conclusion is an hypothesis that provides a causal explanation of the facts stated in the premises, and this explanation is better than any available alternative. The two exceptions are the arguments from the heliocentrism of planetary motions and from the negative correlation of luminosity and mobility. As suggested earlier, this ubiquitous logical structure of the book's argumentation corresponds with two other significant features: that the book provides a confirmation (as distinct from a demonstration) of the earth's motion, and that Galileo could see himself as complying with the ecclesiastical stipulation to limit himself to an hypothetical discussion.

Third, recall that the simplicity arguments for terrestrial rotation (IIA) are all explanatory (inferences to the best explanation), based on the principle of simplicity, and thus probable at best; but one of them deserves to be singled out for its relative novelty and special strength. The argument from the law of revolution claims that daily terrestrial rotation is more likely than universal geocentric revolution because it is simpler, and it is simpler because it is more in accordance with the law of revolution. This law states that the bigger the orbit of revolution, the longer the time required to complete it; and hence, in the geostatic system, although the planets obey this law, the diurnal motion of the celestial sphere contradicts it.

Fourth, Galileo regarded the tidal argument as the strongest, that is, as stronger than any other geokinetic argument. His reason for this judgment is important: as just noted, except for the arguments from heliocentrism and from luminosity (which were obviously weaker), the other geokinetic arguments were hypothetical and explanatory (inferences to the best explanation); in such arguments one has to show that the chosen explanation is better than the alternatives; now, whereas in other cases alternative explanations were available but inferior, for the case of the tidal argument he thought all available geostatic alternatives were physically impossible. However, the same reasoning implies that such an argument, however strong, is not absolutely conclusive, insofar as it is logically, theologically, or metaphysically possible that there might be some alternative explanation; and the divine-omnipotence objection was one way of formulating the latter point.

Fifth, interpreting the tidal argument as such an inference to the best explanation also enables us to formulate a fair and relevant criticism stemming from the later emergence of the Newtonian gravitational explanation of tides. For it would be irrelevant and unfair to fault Galileo's tidal argument by saying that we now know that the true cause of tides is the gravitational attraction of the moon and sun, since this tells us that one of Galileo's key claims is false, but not what is wrong with the reasoning on which he based that claim. However, if we regard Galileo's tidal argument as a basically proper inference to the best explanation, then we can say instead that once Newton formulated his gravitational explanation one would have to take it into account, by comparing its merits with those of the Galilean geokinetic explanation; and then one would be in a position to claim that the latter is inferior to the former.

Sixth, it is obvious that Galileo's case for the earth's motion appeals to a considerable amount of evidence, going far beyond just the consideration of tides. Still, all the fourteen geokinetic arguments listed above are a much smaller set than the geostatic ones; in fact, by one count, there are at least twenty-one of the latter. However, the crucial issue is not that of just counting the numbers of arguments on each side of the controversy, but rather to evaluate their strength; and this was clearly understood and de facto practiced by Galileo. And his evaluation, which is not far from the truth, is that all the geostatic arguments can be effectively criticized, with the exception of the one from annual stellar parallax, which retains some probative weight. On the other hand, the arguments on the geokinetic side all have individually some degree of strength, and together a considerable amount, although admittedly they are not conclusive.

Seventh, more needs to be said about Galileo's evaluation of the arguments on both sides. It is not the case that his criticism of the geostatic arguments and his favorable elaboration of the geokinetic arguments is always completely correct. As we saw in my reconstruction of the book's main argument, sometimes he does not do full justice to his opponents' arguments, and sometimes his own geokinetic arguments contain various errors or do not prove as much as he thinks. For example, in his refutation of the objection from the extruding power of whirling (IIB11), he attempts to prove the mathematical impossibility that the extruding tendency along the tangent (however large it might be) should overcome the downward tendency along the secant (however small it might be); and this is misguided. And in his geokinetic argument from the tides (IVB, IVD), he claims that the earth's motion is the primary cause of the tides; and this claim is substantively false, even though, as just clarified, we must be careful about how this affects the soundness of his reasoning. However, it is not necessary that Galileo's criticism of the con arguments and presentation of the pro arguments be 100% correct. For the confirmation of the earth's motion to succeed, it is sufficient that he be essentially or mostly correct. I believe that this is the case, and that the earlier reconstruction of the book's main argument makes this evident.

Finally, the point just made about essential correctness requires the discussion of at least one major criticism of Galileo's main argument that would affect precisely the issue of essential correctness. The criticism begins by pointing out that the *Dialogue* does not explicitly discuss the Tychonic system of the world. Recall that according to Tycho the five planets revolve in heliocentric orbits, but the sun revolves in a geocentric annual orbit, carrying these planets along with it in the process, while they all also share the universal diurnal motion around the motionless earth. Galileo's neglect of the Tychonic system would represent both logical and methodological flaws.[5]

Logically, the difficulty is that, although Galileo's arguments show the Copernican system to be superior to the Ptolemaic system, they do not show the former to be any better than the Tychonic system. For the Copernican and the Tychonic systems were, supposedly, observationally equivalent and the available evidence could be explained equally well by both; for example, the phases of Venus require only a heliocentric orbit by this planet, but say nothing about whether the annual motion belongs to the earth or to the sun–Venus pair. Here, one logical flaw might be that of attacking a straw man, namely the Ptolemaic system, which by 1632 was generally rejected by astronomers, rather than criticizing the Tychonic system, which was widely accepted and indeed was a serious rival to Copernicanism. Another logical flaw might be the fallacy of affirming the consequent, namely the fallacy of arguing in the form "P is true; if Q then P; therefore, Q is true"; here, Q would stand for the geokinetic thesis, and P would stand for any phenomenon (such as the phases of Venus) which can be explained by means of Copernican causes (see Gingerich 1982).

Methodologically, the neglect of the Tychonic theory would be at least a lapse of objectivity, insofar as objectivity requires the consideration of all available relevant evidence and all available alternative theories. Another methodological flaw might be an appeal to ignorance, insofar as Galileo may be directing his book to a lay audience of non-experts, who are ignorant of the Tychonic alternative, rather than addressing technical astronomers who were acquainted with it.

Despite its popularity and apparent plausibility, such criticism cannot survive critical scrutiny. For although Galileo does not discuss Tycho's alternative explicitly, he does so implicitly, and hence he is not really neglecting it. One reason stems from the fact that the Tychonic theory does, after all, share a crucial common element with the Ptolemaic system; that is, both hold the earth to be motionless at the center of the universe. That is, in both systems, the diurnal motion belongs to the whole universe except the earth, and the annual motion belongs to the sun. Therefore, all the Galilean arguments for the earth's motion and against the geostatic, geocentric thesis undermine Tycho's as well as Ptolemy's world view. This is the case, for example, with the sunspot argument, the tidal argument, and the argument from the law of revolution.

Another reason why Galileo is not really neglecting the Tychonic system is that he does discuss (and refute) the anti-Copernican arguments which had motivated Tycho to devise his compromise system. These are the mechanical objections to the earth's motion criticized in the Second Day of the *Dialogue*, which Tycho had advanced in his *De mundi aetherei recentioribus phaenomenis* (1588) and *Epistolae astronomicae* (1596). Again, the only thing Galileo does not do is to attribute them to Tycho by name.

Another important point to understand is that not every argument supporting some part of the Copernican system necessarily supports the key geokinetic idea, and that Galileo is clearly aware of this. Thus, for example, when he discusses the phases of Venus (IIIB1), he makes it perfectly clear that this phenomenon proves only the heliocentricity of Venus's orbit; that it does not even prove the general heliocentric thesis for all (five) planets; that distinct arguments are needed to support the heliocentric character of the orbits of Mercury, Mars, Jupiter, and Saturn; that, once one has established the general heliocentrism of planetary orbits, a separate and additional argument is needed to decide the issue of whether the annual motion belongs to the earth or the sun; and that the additional argument for the earth's annual motion is based on analogy and so is merely probable, and hence is much weaker than the argument *for* the heliocentrism of planetary motions.

Thus, the objectivity and force of Galileo's critical examination of the arguments on both sides cannot be faulted by claiming that he failed to take into account a third side of the controversy, namely Tycho's. From the point of view of the earth's motion, the Tychonic and Ptolemaic systems are identical, and there is really no such third side. And insofar as Tycho presented relevant objections to the earth's motion, Galileo does discuss them, and even strengthens them before refuting them.

NOTES

1 Newton 1999: 424; cf. Herivel 1965: 35–41, Cohen (1967, 1999: 146).
2 Descartes to Mersenne, 14 August 1634, in Galilei 1890–1909, vol. 16: 124–25; also in Descartes 1897–1913, vol. 1: 303–6.
3 Cf. Galilei (1890–1909, vol. 8: 205; 1974: 162; 2008: 342–43).
4 Cf. Strauss 568 n. 9 (= Sexl-Meyenn 568 n. 9); Pagnini vol. 3: 243 n. 1.
5 See Dreyer 1953: 416, Heilbron 2001: xv–xvi, Koestler 1959: 477, Langford 1966: 123. Cf. Margolis 1991.

9

METHODOLOGY

CRITICAL REASONING AND BALANCED JUDGMENT

9.1 RATIONALITY AND CRITICAL REASONING

One of the most important methodological reflections in the *Dialogue* occurs at the beginning of the Second Day (DML 148–53), in the context of Galileo's summary of the arguments against the earth's motion. This summary, in turn, is given just after his presentation of the simplicity arguments in favor of the earth's daily axial rotation and just before he undertakes the detailed analysis and criticism of each argument against. That methodological reflection elaborates several topics relating to the nature of rationality.

The discussion begins when Galileo mentions the fact that normally the Copernicans are acquainted with the arguments of the Ptolemaics, but the reverse is not true. He regards this as an extremely important difference in their respective procedures, giving the Copernicans a methodological advantage over the Ptolemaics. Although he does not use a label to refer to this particular procedure, we may speak of open-mindedness, and define it generally as the willingness and ability to know and understand the arguments, evidence, and reasons against one's own views. Thus, part of what Galileo is doing in this passage is to reflect on the meaning, nature, value, and role of open-mindedness.

The passage also advances the controversial argument that this difference between Copernicans and Ptolemaics is an indication that the Copernican system is more likely to be true than the Ptolemaic one (cf. Wilkins 1684: 16–17, Westman 2011: 498). In so arguing, Galileo is presupposing the methodological principle that open-mindedness is an indication of the likelihood of an idea, at least in the sense that if one is open-minded one is more likely to arrive at the truth.

This argument was explicitly mentioned in a long list of points of censure compiled by the special commission, appointed by Pope Urban VIII soon after the book's publication, to investigate the many complaints against it.[1] One may not want to join Church authorities in such censure, but the argument should not be accepted uncritically.

What Galileo seems to have in mind is that open-mindedness is valuable partly because it strengthens the view one holds, insofar as one can better defend it from critical objections; moreover, it facilitates the discovery of the truth, which emerges more easily from the clash of opposing arguments. However, open-mindedness may also lead to mental confusion, to not knowing what to believe. This last point introduces a topic which is important in its own right.

In fact, Galileo discusses next a problem which I shall label the problem of misology (namely, hatred of logic) because this label is the one explicitly used in a similar passage in Plato's *Phaedo* (88A-91D); there Socrates tries to dispel the confusions which some of his interlocutors began to feel after hearing several arguments and counterarguments about the immortality of the soul. It is also amazing that Galileo's resolution of the problem is essentially identical to the Socratic one; Socrates argues that the way out of the misologic attitude is to learn, in the words of one translator (Tredennick 1969: 144–45), "the art of logic," i.e., the skill of the "critical understanding" of arguments.

In contrast to misologism, Galileo advances a version of what could be labeled rationalism; for such a label suggests a belief in the power of human reason to arrive at the truth, and he has Simplicio say that there is a one-to-one correspondence between truth and good reasoning on the one hand, and falsity and bad reasoning on the other. It should be noted, however, that the human reason in question is not apriorist speculation but critical reasoning; that is, Simplicio is not saying that one can discover the truth by pure thinking (without experience or observation), but rather that truth can be discovered by collecting all the arguments, evidence, and reasons for and against a given view, and then undertaking a critical analysis of them to determine which are good and which are bad.

However, Simplicio's statement of the correspondence principle is oversimplified in two ways: (1) it gives the impression that the task of distinguishing good from bad arguments is easier than it is, by saying nothing about the fact that arguments may appear good but in reality be bad, and vice versa; and (2) it gives the impression that arguments are simply good or bad, completely valid or completely fallacious, whereas they are usually better or worse, partly correct and partly incorrect, more or less strong. Thus, two qualifications are needed to make this principle viable, because otherwise one could point to the Copernican controversy and Galileo's own *Dialogue* to show that here we have a counter-example to the principle, namely a case where there are good arguments supporting two inconsistent propositions. I believe Galileo would want to qualify Simplicio's statement in precisely these ways, and this may be one reason why the oversimplification is put in Simplicio's mouth, and why at the end of this passage Sagredo makes precisely these qualifications.

Thus, instead of speaking of rationalism, I think it is preferable to speak of rational-mindedness, which may be defined as the willingness and ability to take as the truth (or the probable truth, or the provisional truth, or the acceptable belief) the view supported by the most cogent arguments, the strongest evidence, and the best reasons. Now, in order to be rational-minded, one obviously needs to be able and willing to assess or evaluate arguments, reasons, and evidence; and such evaluation depends on being willing and able to properly analyze and interpret them. The label critical reasoning may be used to refer to such a skill of analyzing, interpreting, and evaluating arguments, evidence, and reasons. Furthermore, in order to be rational-minded, one should also utilize the contrary arguments, reasons, and evidence advanced by one's opponents; that is, we should be able and willing to know and understand them, which is what I labeled open-mindedness.

Then, Galileo may be interpreted as holding that although rational-mindedness is difficult, it is essential in the quest for knowledge; and that it depends crucially on critical reasoning and open-mindedness. In the present passage, he is explicitly reflecting on these methodological ideas. And in the whole book he is constantly practicing them; in this regard, my reconstruction of the book's main argument as a critical defense of Copernicanism (Chapters 4–7) may be taken as a substantiation of Galileo's rational-mindedness, open-mindedness, and critical reasoning.

Finally, there is another important quality of mind displayed by Galileo in the *Dialogue* that relates to the ones just discussed. It may be labeled fair-mindedness, and defined as the willingness and ability to learn from and appreciate contrary arguments, even when one is in the process of criticizing or refuting them. This is something above and beyond open-mindedness, since the latter refers merely to knowing and understanding contrary arguments, whereas fair-mindedness requires that we should be willing to admit their merits, even though we may ultimately reject them.[2]

It is useful to distinguish open-mindedness and fair-mindedness because only the latter notion enables us to understand and appreciate an additional significant feature of Galileo's *Dialogue*. In fact, there are many instances of fair-mindedness in the book.

One of the most instructive examples is Galileo's critique of the anti-Copernican argument based on stellar appearances (IIIB3). This argument claimed that if the earth revolved around the sun, there would be annual variations in the apparent magnitudes and positions of fixed stars, but such variations are not seen. Although Galileo does criticize this argument, his primary response is to clarify exactly what such changes would be.

Another good example is the critique of the argument from the extruding power of whirling (IIB11). Before objecting to this argument in various ways, he strengthens its premise about the existence of such a tendency to extrusion, and he engages in some constructive clarification by correcting the usual misstatements of the argument. And even in the course of his negative criticism, he clarifies the nature of the extrusion that would presumably occur on a rotating earth.

In other cases, Galileo's refutation of the opposing arguments is simple, clear-cut, and decisive, but his statement of what the arguments claim is so clear, eloquent, and incisive that it shows an appreciation greater than that of his anti-Copernican opponents. Such is the case for his discussion of the arguments from the ship's mast experiment (IIB4), and from the behavior and appearance of the planets Mars and Venus (IIIB1).

Another instructive example is provided by Galileo's presentation of the pro-Copernican argument from the annual cycle of sunspot paths (IIIC3); this is an inference to the best explanation, which in the process criticizes alternative geostatic explanations. Galileo points out that Scheiner's geostatic explanation in terms of four solar motions is inferior to the geokinetic explanation in terms of two terrestrial motions and one solar motion; but after this criticism, Galileo adds the hint that in the geostatic explanation the annual precession of the sun's axis of rotation would be as unnecessary as the analogous "third motion" which Copernicus himself (erroneously) attributed to the earth. The hint is cryptic, but we saw earlier that it is sufficient to devise another geostatic explanation with no precession of the solar axis; however, such a revised explanation also turns out to be inferior to the geokinetic one, because it is refuted by the non-existence of a daily cycle of changes in sunspot paths. This hint shows that Galileo is able and willing to explore the hidden strengths of a contrary view, even when hidden to its proponents.

Such fair-minded practice in the *Dialogue* is ubiquitous, clear, and unmistakable. Galileo shows some reflective awareness of it when, at the beginning of the Second Day, in the passage mentioned above, he states that "the followers of the new system produce against themselves observations, experiments, and reasons much stronger than those produced by Aristotle, Ptolemy, and other opponents of the same conclusions" (FIN 147–48; cf. DML

148). A more explicit and memorable statement was formulated by Galileo on another occasion, during the trial proceedings of 1633, when he claimed that in the *Dialogue* he had abided by the following principle: "when one presents arguments for the opposite side with the intention of confuting them, they must be explained in the fairest way and not be made out of straw to the disadvantage of the opponent."[3]

Thus, in this passage at the beginning of the Second Day, Galileo is reflecting on the nature and interrelationships among open-mindedness, rational-mindedness, critical reasoning, and fair-mindedness, and their role in the search for truth. He tries to illustrate and clarify these concepts, and to formulate and justify a number of methodological principles corresponding to them. All these reflections arise naturally and directly in the context of the Copernican controversy. They represent Galileo's response to the fact that for a long time during the Copernican revolution reasonable people could take opposite sides on the issues by presenting arguments, reasons, and evidence seeming to favor conflicting conclusions.

Galileo's response is also a plea to avoid both dogmatic one-sidedness and skeptical despair, both anti-intellectual misologism and rationalistic apriorism, both simple-minded convergence and simple-minded divergence between physical truth and human reasoning, and both acquiescence in and dismissal of opposing views. And in turn, this Galilean plea is a good illustration of an attitude or quality of mind which may be called judgment or judiciousness (or judicious-mindedness), and which may be defined as the avoidance of one-sidedness and of extremes. As we shall see, judgment (in the sense of judiciousness), together with rationality or reasoning (in the sense of open-mindedness, rational-mindedness, critical reasoning, and fair-mindedness), turns out to be the most fundamental trait of Galileo's methodological reflections.

9.2 AUTHORITY AND INDEPENDENT-MINDEDNESS

Galileo also had to face the fact that the geokinetic world view contradicted that of the most highly respected intellectual authority on the subject—Aristotle. This was especially problematic because

Galileo did not want to reject all use of authority in general and of Aristotle's authority in particular. Such total rejection would have been too extreme, as it were. Several passages discuss his response to this situation.

The longest and most sustained of these discussions occurs at the very beginning of the Second Day (section IIA, DML 123–32). Negatively viewed, it is a critique of Aristotelian authority, i.e., a criticism of the misuse and abuse of the authority of Aristotle; positively speaking, it is a plea for independent-mindedness.

It begins with the portrayal of an Aristotelian who witnesses the anatomical dissection of a human cadaver; is shown that the nerves originate in the brain rather than the heart; and comments that he would have to believe what he saw were it not for the fact that Aristotle had said that the nerves originate in the heart. This memorable image is an example of an uncritical and excessive subservience to authority; we have no methodological argument but rather a vivid illustration of an undesirable methodological trait. The same is true for the equally memorable portrayal of another Aristotelian's reaction to the first news of the invention of the telescope: he claimed that Aristotle anticipated the invention in the passage which describes the phenomenon of the stars becoming visible during daylight if observed from the bottom of a well; the well was analogous to the tube holding the telescope's lenses, and analogous was also the effect of being able to see something that cannot be seen without such apparatus.

At the opposite extreme, there is the fictionalized story of a scholar who had written a book on the soul claiming, on the basis of certain passages, that Aristotle denied its immortality. When told that such a book would not receive permission to be published (because of the theologically heterodox character of this conclusion), he decided to elaborate instead the interpretation that Aristotle held the soul to be immortal, based on other passages. Such opportunism is a memorable example of an excessive and undesirable disregard for the words and texts of an authority.

Thus, for Galileo the proper use of Aristotle's authority involves a judicious independence of mind, as contrasted to both uncritical subservience and the arrogant self-centeredness of thinking one can make the authority stand for whatever one wishes. When so stated, this Galilean methodological rule is hard to disagree with. As with all cases of judgment calls, however, the real challenge is putting it into practice.

This passage also contains a more specific suggestion. That is, the proper use of authority involves focusing on its reasoning; on the reasons underlying the authority's assertions, and not its mere assertions; on understanding why it said what it did. In other words, we need to understand, analyze, and evaluate the authority's arguments; the proper use of authority involves critical reasoning in a crucial way.

In accordance with this principle, other passages provide at least three examples of critical reasoning vis-à-vis authority. In the discussion of the geocentric argument from natural motion (IA4), Simplicio does properly present Aristotle's argument rather than his mere conclusion (DML 38–39). However, the discussion then moves to the evaluation of this argument, and the issue becomes whether authority can play a role at that level.

This issue arises in an especially natural way because Aristotle also happened to be the founder of the science of logic, and so the possibility arises that one may use his logical authority in the defense of the favorable evaluation of the geocentric argument from natural motion. Thus, when Salviati faults the argument as question-begging, is it proper to reply that this argument cannot have this flaw because it was advanced by the founder of the science of logic? Galileo's resolution of this problem is to distinguish between logical theory and logical practice; equate the science of logic with logical theory, i.e., the theory of reasoning; equate argumentation with logical practice, i.e., the practice of reasoning; and then say that Aristotle's authority as a logical theorist cannot be used in the present context, which is one of concrete reasoning. The general methodological lesson here is that when in an appeal to authority we try to utilize the authority's reasoning, we cannot also utilize the authority's own evaluation of it, which must rather be always regarded as an open question; this is true even when the given authority happens to be also a genuine authority in the general theoretical evaluation of arguments.

In other cases, appeal to authority is subject to other principles. Consider, for example, the discussion of the center of the universe at the beginning on the section of the heliocentrism of planetary revolutions (IIIC1). Clearly, it is insufficient for an Aristotelian to say merely that the earth is at the center of the universe because Aristotle said so; one must also report Aristotle's reasons. One of these involved the argument that the earth is at the center of the universe because it is at the center of planetary revolutions. Such an argument presupposes the idea that the center of the universe is the center of planetary revolutions; this may be regarded as the Aristotelian definition of the center of the universe.

Now, Galileo thought that in his own time the new telescopic evidence was such that the minor premise of this argument could be conclusively refuted, and one could prove instead that the sun is at the center of planetary revolutions. It followed that if one applied the Aristotelian definition to the new situation, one would have to conclude that the sun is at the center of the universe. The point is that, if and to the extent that the reason for Aristotle's belief in universal geocentrism was his belief in the geocentrism of planetary revolutions, then insofar as one can show that planetary revolutions are heliocentric, this fact should become a reason for believing in universal heliocentrism instead. The general methodological lesson is that appeal to an authority's reasoning opens up the possibility that, when new premises displace the old ones, the authority's own manner of reasoning will lead to different conclusions. That is, appeal to the argument of a past authority to help decide a current issue naturally leads to considering what conclusion that authority would reach in the current circumstances by reasoning in the same manner followed previously.

The same principle is illustrated by Aristotle's observational argument for heavenly unchangeability (IA6), except that the situation there is complicated by the fact that the new different conclusion is also justified by other Aristotelian assertions. That is, at first we have Aristotle's argument that the heavens are unchangeable because no heavenly changes have ever been observed; this argument presupposes that normal observation corresponds to reality; in this case, the authority's manner of reasoning is simply to base the conclusion on observation. Now, when observation reveals instead that the heavens are changeable, as it happened at the time of Galileo, then the Aristotelian manner of reasoning leads to the un-Aristotelian conclusion that the heavens are changeable. Thus, following the authority with regard to reasoning implies disregarding it with regard to the particular physical conclusion in question. It is obvious that the detection and resolution of such a tension requires both critical reasoning and the skill of discernment which is involved in judgment.

However, here the tension also emerges in another way. For Aristotle also explicitly asserted that in the search for truth, sensory observation should be given priority over any intellectual theorizing; this is a general methodological principle which he asserted and used on many occasions. Thus, in exploring whether the heavens are changeable, one might use Aristotle's authority by appealing to this principle. Now, an application of this principle at the time of Galileo would yield that observations reveal that the heavens are changeable. By Aristotle's own principle, these observations should be given priority over his own speculations to the contrary. Hence, in this case if we accept Aristotle's *methodological* authority, we are led to reject his *scientific* authority with regard to a particular physical theory. In Galileo's own memorable words: "Consider now these two propositions, which are both parts of Aristotle's doctrine: the last one, claiming that one must put the senses before theorizing; and the earlier one, which regards the heavens as unchangeable. The last one is much more solid and serious than the earlier one, and so it is more in accordance with Aristotle to philosophize by saying 'the heavens are changeable because so the senses show me' than if you say 'the heavens are unchangeable because theorizing so persuaded Aristotle'" (FIN 104; cf. DML 63).

In short, in some cases, appeal to authority is rendered problematic by the fact that the authority's texts contain divergent elements, and so independence of mind (together with the judgment and critical reasoning that go along with it) is required.

9.3 SENSORY OBSERVATION AND THEORETICAL REASON

One of the most fundamental problems in inquiry is that of the nature of observation, its relationship to thinking, and its role in the search for truth. Here, observation is only one of a cluster of terms and activities which includes the senses, sense experience or sensory experience, perception, and experiment. Further, the label empirical may be used to refer neutrally to anything pertaining to these, while empiricism may be taken to mean an excessive emphasis on them. On the other hand, thinking belongs to a cluster which also includes the intellect, intellectual activity, ideas, theorizing, speculation, conceptualization, reason, and reasoning; the label conceptual may be taken to be the counterpart of empirical, and apriorism or intellectualism the counterpart of empiricism.

In Galileo's work in general, and in the *Dialogue* in particular, there are indications of an inclination toward an empiricist approach, as well as indications of an inclination toward an apriorist approach. Accordingly, ignoring such complexity, some scholars have portrayed him simply as an empiricist (e.g., Drake 1978, 1990), and some have viewed him simply as an apriorist (e.g., Koyré 1966, 1978).

However, both interpretations are injudicious exaggerations of things Galileo said or did. Nor is it proper to claim that he was simply inconsistent in his methodological reflections and practices. He was not inconsistent vis-à-vis empiricism and apriorism, because he did not universalize his reflections and procedures in the way required to generate an inconsistency. Nor do we have a case of the otherwise common divergence between words and deeds, because when his methodological pronouncements are examined in context they can be seen to correspond to his deeds. That is, Galileo was a critical empiricist, or a critical apriorist, which is to say neither a real empiricist nor a real apriorist; instead he was a judicious practitioner of both the empirical (not empiricist) approach and the conceptual (not apriorist) approach, and this judiciousness may be detected in both his practice and his reflections.

One of the most instructive illustrations involves the fundamental principle of empiricism, which stipulates that observation has priority over theorizing. This principle is explicitly (and accurately) attributed to Aristotle in the discussion of the geocentric argument from natural motion (IA4) and in the discussion of the observation of heavenly changes (IA6). Moreover, the principle is explicitly endorsed by Galileo when in the latter discussion he calls this stipulation "fitting" (FIN 96; cf. DML 57), and it is used by him in the same discussion when he argues that the heavens are changeable because heavenly changes can now be observed. Furthermore, Galileo's attitude toward Copernicanism before the telescopic observations is in accordance with this principle, as he admits in the discussion of the objections based on the behavior and appearance of Mars, Venus, and the moon (IIIB1). This admission occurs when he expresses amazement at the fact that Aristarchus and Copernicus were able to do the reverse (i.e., give priority to their geokinetic idea over the observational, especially astronomical, counterevidence); and when he says that without the telescope he would have been unable to join them (DML 381).

On the other hand, there are also indications of an opposite Galilean inclination, namely toward apriorism. For example, in the discussion of the ship's mast experiment (IIB4), one of the issues is what is the true result of the experiment. It is certainly striking that in the *Dialogue* Galileo does not resolve the issue by actually doing the experiment or reporting the result of some actual experiment; instead he gives a plausible theoretical argument showing that the rock will fall to the foot of the mast.

Or consider the discussion of the apparent positions and magnitudes of fixed stars (IIIB3). There Galileo certainly admits both that the Copernican theory implies the existence of some annual changes in stellar appearances, and that observation (even telescopic observation) reveals none. However, he does not thereby abandon the theory, and instead chooses to give priority to the theory by disregarding available observations and explaining them away as due to various causes such as immense stellar distances, lack of systematic observation, and lack of sufficiently powerful instruments.

Finally, recall the geokinetic explanation of the diurnal period of tides (IVB). There, after elaborating his account of the primary cause, Galileo admits that perhaps it conflicts with the observation of a six-hour interval between high and low tide. However, he holds on to the theory and explains away the observation as due to other concomitant secondary and tertiary causes, distinct from the primary cause (DML 502). His explanation may be deemed plausible, but from the point of view of the methodology of theory versus observation the difficulty remains.

The general difficulty remains in the sense that on this occasion (and in the similar cases just mentioned) Galileo seems to be giving priority to theory over observation, whereas in other cases (such as those mentioned earlier) he does the reverse; thus, in regard to the issue of priority of theory or observation, he does not seem to always follow the same principle. Moreover, it should be noted that in those cases where Galileo disregards observations, he explains them away. However, to use this locution (explaining away) is prejudicial and one-sided; such language presupposes that one is taking the side of observation, in the sense of simple-minded empiricism; so it is preferable to speak simply of explaining available observations, which is perfectly legitimate. To take such legitimate explanation into account complicates the situation, but also suggests that a more sophisticated principle is needed to make sense of such behavior, and that the fundamental principles of empiricism and apriorism are both oversimplifications.

For example, one might hold generally that neither is observation prior to theory, nor is theory prior over observation, but rather sometime the former holds and sometimes the latter, depending on certain conditions. That is, when there is a conflict between theory and observation, the theory should be abandoned if the conflict cannot be explained in any other way (for example, heavenly unchangeability and sunspots); but the theory need not be abandoned if the conflict can be explained as due to other causes (for example, earth's motion and no detection of stellar parallax). Galileo comes very close to explicitly stating such a principle in the discussion of the apparent positions and magnitudes of stars (IIIB3, DML 448–50).

Similarly, Galileo's conceptual orientation with regard to the ship experiment can be taken, not as a sign of apriorism, but rather of the fact that sometimes one can give a justification of a truth based on other more easily accessible premises, especially if one has already observed it empirically. In fact, as I mentioned earlier (IIB4), and as Galileo reported in his "Reply to Ingoli," he had made this experiment and obtained an anti-Aristotelian result.[4] This would give priority to observation chronologically speaking, or in the context of discovery; but theory would receive priority logically speaking, or in the context of justification. In the discussion of the observation of heavenly changes (IA6), Galileo himself applies this distinction to the case of the Aristotelian thesis of heavenly unchangeability, when he claims that probably Aristotle first observed a lack of heavenly changes and later formulated the theoretical justification based on the connection between change and contrariety (DML 58). Therefore, to apply the distinction to Galileo's own conceptual approach to the ship experiment is in the spirit of his own methodological reflections.

Let us look at another important illustration of the methodological problem of the role of observation, which was crucial in the Copernican controversy. This is the principle that sensory experience is normally reliable, namely that normally the human senses tell us the truth. This principle is a species of empiricism, in the sense that it expresses a way of emphasizing observation and gives observation special importance in the search for truth. We have already seen that Copernicanism contradicted this principle, insofar as the earth's motion cannot be directly seen, perceived, felt, or sensed. The so-called objection for the deception of the senses was a formalization of this difficulty (IIC3).

The principle of the reliability of the senses does not deny the existence of perceptual illusions. For example, everyone knew that a straight stick half immersed in water appears bent, and that the shore appears to move away from us when we look at it while standing on a boat which is moving away from the shore. However, these deceptions were accidental, temporary, and easily corrected by means of further observations, whereas the sensation of the earth being at rest had none of these properties. Thus, the known perceptual illusions did not present a problem for the principle, whereas the earth's motion did.

Moreover, the principle did not deny the existence of exceptional cases when the senses were not reliable because they were diseased, damaged, or disturbed by external factors such as wine or drugs. However, these were exceptions to the norm of healthy, natural sense-experience. On the other hand, the earth's motion was not observable under normal conditions.

Galileo confronts the problem head on (IIC3), in the sense that all his replies involve a partial abandonment of the original principle and a willingness to diverge from this kind of naïve empiricism. For example, he argues that experience with navigation shows that normally we can only perceive changes of motion, and not mere motion; thus, the earth's motion is not something susceptible of being perceived by normal observation. While this suggests that Copernicanism does not involve a deception of the senses, it also suggests that the principle of the normal reliability of the senses is not true.

Galileo also argues that, insofar as there is a deception, it may be more proper to speak of a deception of reason than of the senses. For the principle of the reliability of the senses is actually a principle of reasoning which validates inferences from what is revealed by normal observations to what exists in physical reality. Now, if Copernicanism is right, that means that there are cases where something is revealed to normal observation (the earth at rest) but does not correspond to reality. This means that such inferences are not always correct; therefore, this general principle of inference is invalid and must be qualified. Still, the qualifications involve our reasoning, not our senses.

Finally, insofar as the senses are unreliable, that does not mean that they are totally unreliable, that they never tell us the truth. One must not go from one extreme (the senses are normally reliable) to the opposite extreme (the senses are normally unreliable). Rather, normal observation is sometimes reliable and sometimes unreliable. We must learn to live and to operate with this more qualified principle, which requires more of an exercise of our judgment.

Once again, what emerges more than anything else in Galileo's methodology is his interpretation of the situation in terms of critical reasoning on the one hand, and his judiciousness on the other. The same is suggested by one last and important illustration of the methodology of theory and observation—the telescope.

Galileo's readiness to make scientific use of such an artificial instrument indicated that he wanted to pursue the empirical method to a deeper level than many of his empiricist opponents. For telescopic observation is a special case of observation, and empowering the human senses with artificial instruments is still a case of using the senses.

On the other hand, such an artificial instrument is essentially a product of human theorizing. Galileo may not have built his first telescope by a systematic application of optical theory, as he suggests in some of his writings; he may have followed a more practical, trial-and-error procedure. Nevertheless, the latter as well as the former procedure involve significant thinking that precedes the use of the eye in the new context. Thus, telescopic observation is observation guided by reason in an additional sense, as compared to normal, natural observation.

In short, observation with the help of artificial instruments is an especially telling example of the judicious combination of the senses and the intellect, of how the element of reasoning is present (presupposed) in an activity (observation) that at first seems very different from it.

9.4 MATHEMATICAL TRUTHS AND PHYSICAL REALITY

Mathematical ideas play such a central role in modern physical science that there can be no doubt that, as Galileo expressed it, "to want to treat physical questions without geometry is to attempt to do the impossible" (FIN 193; cf. DML 236). Yet, as the term—geometry—used by him to refer to mathematics reminds us, mathematics is not a static entity, and since Galileo's time it has come to subsume more and more new ideas that were inconceivable to him and his contemporaries. Actually, even in Galileo's own work, mathematics frequently involved quantitative considerations which he himself would have subsumed under arithmetic, and distinguished from geometry as narrowly conceived; still, as previously noted, he frequently used the terms geometry and mathematics interchangeably.

Not only is mathematics a dynamic discipline, but its development has been intimately interwoven with that of physical science. Often physical problems have forced scientists to invent new branches of mathematics, as happened soon after Galileo with the development of the calculus in order to come to terms with the phenomenon of motion. Sometimes the interaction has gone in the other direction; that is, the availability of some independently developed branch of mathematics has led to the understanding of certain natural phenomena and the creation of new branches of physical science.

It follows from these considerations that the methodological problem of the role of mathematics must be taken to include the question, "What is mathematics?" Now, some commentators tend to equate mathematical thinking with apriorist thought, since mathematics is an a priori science and observation has no role in mathematical inquiry. Given the nonempirical character of mathematics, this is a plausible conception. However, if by mathematics we mean primarily its apriorist aspect, then the methodological problem of the role of mathematics becomes indistinguishable from the previous problem of the role of observation, for both would be special cases of the problem of the relationship between empirical and conceptual elements in the search for truth; then, there would be little new to be added to the discussion in the preceding section.

However, by mathematics Galileo meant primarily deductive reasoning involving abstract entities such as numbers and geometrical figures. By deductive reasoning, he meant reasoning where the conclusion follows necessarily from the premises, so that if the premises are true then the conclusion must be true, and it is impossible for the premises to be true and the conclusion false (DML 118–21, 236–41). This conception is the relevant one if we want to discuss the methodology of the role of mathematics as a distinct problem. However, although we can formulate such a distinct problem, with regard to the question of the proper interpretation of Galileo's own text and methodology, we can draw a parallel with the difficulties discussed earlier with regard to the problem of the role of observation.

To see this, let us go back to Galileo's claim that "to want to treat physical questions without geometry is to attempt to do the impossible" (FIN 193; cf. DML 236). Although this claim is indisputable today (since we are in a sense inheritors of the science of Galileo), it was relatively controversial at his time. However, this Galilean reflection is not an isolated example of his commitment to the mathematical approach. Another formulation is worth mentioning, which appears not in the *Dialogue* but in *The Assayer*:

this is the eloquent image that the book of nature is written in mathematical language.[5]

Based on such remarks, some readers (e.g., Koyré 1966, 1978; Shea 1972) attribute to Galileo a mathematicist methodology which is tantamount to an obsession with mathematics and the power of the mathematical approach in the study of nature. I label such extremist methodology mathematicism and distinguish it from the mathematical approach as such, which may or may not involve such an excessive emphasis and universalization. This is analogous to the distinctions between empiricism and the empirical approach, and between apriorism and the conceptual approach.

The distinction can be illustrated in terms of the Galilean reflection quoted above. It could be construed literally to mean that mathematical considerations are necessary in the treatment of many physical questions. When so construed, it expresses a moderate, judicious commitment to the mathematical approach in natural science. However, it could be taken to mean that mathematical considerations are both necessary and sufficient in the treatment of all aspects of all physical questions. On this second interpretation, the claim would be a good statement of mathematicism. I believe this second, mathematicist, construal is an exaggeration and universalization on the part of the reader, and not Galileo's position.

The mathematicist interpretation also involves taking the sentence out of context. There are several levels here. The most immediate one is that the sentence appears (in section IIB11) at the end of Galileo's attempt to prove the mathematical impossibility of extrusion on a rotating earth (DML 229–36). That is, he first explains that on a rotating earth, bodies would have a tendency toward extrusion along the tangent due to the rotation, as well as a downward tendency along the secant due to gravity, and then he tries to show that the downward tendency (however small) would always be able to overcome the extruding tendency (however large). He tries to prove this claim on the basis of the geometry of the situation in the neighborhood of the point of contact between a circle and a tangent, and the behavior of the external segments (called exsecants) of the secants drawn from the center of the circle to the tangent. Now, the important point to note here is that this proof consists not of a single mathematical deduction but of three distinct considerations. This is important because one mathematical proof should have been sufficient, since in valid mathematical reasoning the conclusion follows necessarily from the premises; so, if he is giving three proofs, perhaps he is not sure about their validity or applicability. This suggests that the power of mathematics may not be as great as suggested by the mathematicist interpretation of this Galilean sentence.

At the next level of context, it is important to stress that the sentence is uttered in the middle of the critique of the anti-Copernican objection from the extruding power of whirling. Galileo's substantial criticism of this objection is that it is rendered quantitatively invalid by its neglect of relevant quantitative considerations. One of these considerations is the one we have just discussed, involving the (problematic) mathematical impossibility of extrusion. Another consideration, indirectly suggested in the text just preceding it (DML 226–27), is that the relevant physical parameters (terrestrial radius, rotating speed, acceleration in free fall) are such that the downward tendency happens to exceed the extruding tendency. And a third consideration at the end of the passage (DML 245–53) is that the cause of extrusion increases with the linear speed but decreases with the radius, and so it would really be very small due to the fact that the linear speed at the equator is very small compared with the earth's radius. Thus, Galileo's main point is that without mathematical considerations of this sort it would be impossible to understand why the extrusion objection is ineffective.

However, this does not apply to every aspect of the same problem. For example, the first thing Galileo explains (DML 218–20) is that the extrusion objection as usually formulated is improperly stated, and so it must be reformulated to see that there is a problem at all. This is a matter that involves a constructive clarification and a logical criticism of the argument, and no mathematical considerations are involved.

Moreover, in the immediate context where the quoted remark is made, Galileo embarks on a long methodological discussion of the application of mathematical truths to physical reality (DML

236–44). Unless this discussion is to be taken as an irrelevant digression, its function must be to express some qualifications against a possible mathematicist misinterpretation of his mathematical approach. The complex, nuanced, and sophisticated nature of the content of these methodological reflections about the relationship of mathematics and physics constitute the best evidence of Galileo's mathematical (but nonmathematicist) methodology.

These reflections may be reconstructed as follows. First, mathematical truths are about abstract entities in the sense that they are statements about the necessary consequences of certain definitions and axioms; for example, the proposition that a sphere touches a plane at a single point is about abstract spheres and planes insofar as it is a necessary consequence of the definition of a sphere and the axiom that a straight line is the shortest distance between two points in a plane. Second, mathematical truths are also about physical reality, although only conditionally; that is, a mathematical proposition is physically true if and only if the abstract entities to which it refers happen to exist as material entities in physical reality; for example, the proposition about spheres and planes is physically true in the sense that if there happen to be material spheres and planes then they touch in only one point. Third, mathematical truths are applicable to physical reality because and insofar as material entities instantiate or approximate abstract ones, for when material entities do not approximate one type of abstract entity they are bound to approximate another type; for example, if and to the extent that a material sphere touches a material plane in more than one point, they instantiate abstract spheres and planes that are imperfect, and for these it is equally true in mathematics that they touch at more than one point. Finally, the real challenge is to find the proper type of abstract entity in terms of which to interpret physical phenomena; although one can be sure that the latter must correspond to some type of abstract entity treatable mathematically, one cannot be sure of which one; for example, this may be the difficulty with the relevance of the earlier mathematical analysis of extrusion.

At a more general level of context, we can point out that the *Dialogue* contains many other critiques of anti-Copernican objections, and most of these do not involve mathematical considerations. Indeed, the book is largely qualitative and thus belies the mathematicist interpretation of Galileo's commitment to mathematics. What this means is that in order to attribute mathematicism to Galileo, one must also attribute to him the proverbial inconsistency between theory and practice, namely that his words express mathematicist pronouncements, but his deeds do not live up to them.

However, I think it is more proper to construe the remark literally rather than to exaggerate it out of proportion; to consider it in context; and to make it correspond to what he did. When all this is done, the remark becomes evidence of a moderate, judicious commitment to the mathematical (rather than mathematicist) approach.

There are other important indications of Galileo's judiciousness about mathematics. An especially important one involves the discussion with which the *Dialogue* begins (IA1). The very first thing he does in the book is to dissociate himself from various abuses and misuses of the mathematical approach popular among the followers of Pythagoras; the key practice he wants to reject is the confusion of the world of numbers with the world of nature. He wants to express his commitment to the mathematical approach, but to ensure that this is not misunderstood.

Finally, there are other methodological problems and principles discussed in the *Dialogue* besides those reconstructed so far. However, these other topics are either intrinsically less important, or less extensively discussed in the book, or already incorporated into the reconstruction of the book's main argument. Thus, we need not examine them here, but can merely mention them: the nature and role of hypotheses, the powers and limitations of the human mind, the principle of simplicity, teleology and anthropocentrism, Holy Scripture, and divine omnipotence.

NOTES

1 Galilei (1890–1909, vol. 19: 327; 2008: 275), Finocchiaro 1989: 222.
2 Here the terms fair-mindedness and open-mindedness do not have a moral connotation, but merely methodological import; cf. Ennis 1996, Fisher 1991,

Govier 1999: 155–80, Johnson 2000: 143–79, Scriven 1976: 166–67, Woods 1996: 650–62.

3 In Finocchiaro 1989: 278, Galilei 2008: 283; cf. Galilei 1890–1909, vol. 19: 343. For a clarification of the problematic context of this remark, see Finocchiaro 2011: 61–62.

4 In Galilei 1890–1909, vol. 5: 545; Finocchiaro 1989: 184. Cf. Conti 1990: 230–31.

5 Galilei (1890–1909, vol. 6: 232; 2008: 183); Drake and O'Malley 1960: 183–84. Cf. Biagioli 1993: 306–7.

10

RHETORIC
PERSUASION AND ELOQUENCE

We have already seen (Chapter 8.2) that Descartes expressed a favorable judgment on the probative strength of the *Dialogue*'s main argument and on its contribution to the science of motion. However, the judgment we quoted earlier is followed by other remarks that criticize the book from the point of view of its persuasiveness. In fact, the relevant passage states that Galileo "philosophizes pretty well about motion ... His reasons proving the earth's motion are very good; but it seems to me that he does not present them as one must in order to be persuasive, for the digressions which he intermingles in their midst have the result that one no longer remembers the first by the time one reads the last."[1]

I believe this Cartesian criticism is insightful, fair, and well founded: there are digressions in the *Dialogue*, and they do diminish the otherwise considerable persuasive force of Galileo's confirmation of the earth's motion. In fact, I would add several others features that detract from the persuasiveness of its main argument. For example, there is the book's dialogue form and consequent dramatic interplay among the three speakers (Salviati, Sagredo, and Simplicio); this can cause confusion to the reader. There is also the literary style of Galileo's prose, which is anything but prosaic, but rather full of eloquent expressions and aesthetically pleasing passages; this may generate distraction for the reader. And there is the practical context stemming from Galileo's struggle with the Church, which is always in the background and occasionally looms into view; this indicates that Galileo does not always mean what he says, or say what he means, thus making readers uncertain about whether they understand the text.

Because of such difficulties, my reconstruction of the main argument (Chapters 4–7) discounted, ignored, or minimized these and other similar features of the book. However, it is now time to pay attention to them. They may be subsumed under the general notion of "rhetoric," even though this term is ambiguous, and so discussions of a book's rhetorical content, form, and significance tend to focus on many disparate things.[2] Thus, our first task will be to bring some order into these meanings.

10.1 THREE SENSES OF RHETORIC

Scientists tend to have a certain aversion toward rhetoric, for there is a tradition of anti-rhetoric in science. On this point, as for so many other issues about the nature of scientific inquiry, the *Dialogue* provides classic illustrations. For example, in the middle of the passage on heavenly changes (IA6), there is a discussion of the nature of sunspots where the issue is whether they are part of the body of the sun like clouds on the earth, or swarms of planets revolving around the sun and partially obscuring our view from the earth when a number of them get in our line of sight. The latter was the Aristotelian interpretation, which retained the doctrine of the unchangeability of the heavenly bodies, whereas the former was Galileo's (correct) view. Toward the end of this discussion, Simplicio expresses his confidence that, even if the planetary interpretation should turn out to be incorrect, others will be able to come up with a theory compatible with heavenly unchangeability. To this Salviati replies:

If what is being discussed were a point of law or of other human studies in which there is neither truth nor falsehood, one could have great confidence in intellectual subtlety, verbal fluency, and superior writing ability, and hope that whoever excels in these qualities could make his reasoning appear and be judged better. But in the natural sciences, whose conclusions are true, necessary, and independent of the human will, one must take care not to engage in the defense of falsehood because a thousand Demostheneses and a thousand Aristotles would be overcome by any average intellect who may have been fortunate enough to have discovered the truth. Therefore, Simplicio, give up your thought and your hope that there could be men so much more knowledgeable, learned, and well-read than we are as to be able to turn truth into falsehood, against nature.

(FIN 101; cf. DML 61)

And in the margin of the page, Galileo writes the comment that "in the natural sciences the art of oratory is ineffective" (DML 61). Although he does not use the word "rhetoric" here, the connection with rhetoric is unmistakable because of his reference to Demosthenes, his talk of "the art of oratory," and the fact that a contemporary critic (Antonio Rocco) explicitly used the term "rhetoric" in his comments (FAV 628–29).

It is obvious, however, that such a methodological pronouncement cannot be taken at face value. Indeed, practicing scientists are not always the best interpreters of their own activities, and their words and their deeds with regard to scientific method do not always correspond. Now, although this does not mean that their words and deeds never correspond, and although for the case of Galileo I have argued (Chapter 9) that they usually do correspond, here we seem to have an exception to this Galilean rule; for there is overwhelming evidence that he was a constant practitioner of rhetoric. However, these are not the points I wish to stress at the moment. Rather, the main purpose of the above quotation is to introduce a number of relevant meanings of the notion of rhetoric.

One of these is a meaning which Galileo does not explicitly have in mind, but which he is implicitly using and which also corresponds to a common present-day conception. That is, in making the anti-rhetorical pronouncement just mentioned, and in other passages where he expresses a similar sentiment, Galileo is himself engaged in a certain type of rhetoric: he is attempting to convey a particular impression. The fact that the content of this message is "anti-rhetorical" is not relevant at the moment. The point is that there are many occasions when scientists want to project a certain image, and it is important to recognize that they are engaged in such image-projection, and that such an image does not necessarily correspond to what they are actually doing, although the fact that they cherish such an image may be itself important.

This is the sense of rhetoric in which one speaks of "campaign rhetoric" in politics, when candidates for office make statements meant to create certain impressions in the minds of the voters. This sense of rhetoric also corresponds to the domain of public relations, in which an organization tries to create a favorable impression with the public at large. However, here my main point is that in this passage Galileo expresses a rhetoric of anti-rhetoric, and that this is a typical attitude in modern science.

However, if the rhetoric of anti-rhetoric is to avoid being a contradiction in terms, the sense of the two occurrences of the term "rhetoric" must be different. That is, the rhetoric which is being explicitly criticized in this type of rhetoric must be different from the rhetoric which is expressing the criticism. In fact, at least part of what Galileo is against in this passage is oratory such as that in which Demosthenes excelled, i.e., the art of eloquent expression. This is obviously a second common meaning of the term, and one of our present problems is whether it is indeed true that rhetoric, in the sense of verbal eloquence, has no function in scientific inquiry.

Actually, this Galilean passage suggests a third sense of rhetoric. For, besides expressing his misgivings about the scientific effectiveness of "verbal fluency", "superior writing ability", and Demosthenes's type of talent, the passage expresses misgivings about the type of intellectual subtlety in which Aristotle excelled. Although it is not completely clear what Galileo is referring to, let us take this to be referring to persuasive argumentation, to be elaborated below.

There are, then, three senses of rhetoric that are relevant for the full appreciation of a great book like the *Dialogue*. One of them is only implicitly used by Galileo in this passage, while two others are explicitly mentioned by him there. They pertain, respectively, to the arts of effective communication, eloquent expression, and persuasive argumentation.[3]

10.2 "MERE" RHETORIC

Let us begin with the first sense of rhetoric: the projection of images, the production of impressions, the communication of messages. Far from having no role in scientific inquiry, this type of rhetoric is unavoidable in science simply because science is a social enterprise practiced by human beings and addressed to other human beings. Even if science were an entity completely independent of the knowing subject, there would be some need for a corresponding rhetoric, i.e., to project the image that images are irrelevant.

However, the important point for this type of rhetoric is not its pervasiveness, but rather the fact that one must learn to see beneath, above, and beyond it in order to appreciate the more important things that are going on. The reason for this stems from the very definition of this type of rhetoric; it involves appearances, impressions, images, and messages sent by contrast to reality, substance, actuality, and messages received. In fact, calling certain things rhetoric in accordance with the present meaning is a way of diminishing their importance.

On the other hand, to say that such rhetoric is relatively unimportant is not to say that it is completely unimportant. Usually it will have a purpose or function, for even appearances have their reality. The mistake is to take such rhetoric at face value—to mistake the appearance for reality. Let us look at some other examples.

The actual content of the *Dialogue* is, as we have seen (Chapter 8), a significant confirmation of the geokinetic hypothesis, grounded on a critical examination of the arguments on both sides and on a synthesis of astronomy, physics, mathematics, and philosophy. By confirmation is meant an argument showing that the geokinetic theory is more likely to be true than the geostatic theory—that the former should be accepted in preference to the latter. This is to be contrasted with a process of conclusively proving that the geokinetic view is absolutely true.

Despite this effective content, there is in the book a considerable amount of rhetoric trying to convey different impressions. To understand this rhetoric, it is important to recall the book's historical background. A crucial point is that the work was written at a time when, as a Catholic, Galileo was bound by at least a decree issued by the Church in 1616 to the effect that one could neither hold nor defend the geokinetic thesis; and we have also seen that perhaps he was also bound by more stringent restrictions about the kind of discussion of this topic (hypothetical or realist) he could engage in.

One of these rhetorical impressions is conveyed by the full wording of the long title, which translates as follows: "Dialogue by Galileo Galilei, Lincean Academician, Extraordinary Mathematician at the University of Pisa, and Philosopher and Chief Mathematician to the Most Serene Grand Duke of Tuscany; where in meetings over the course of four days one discusses the Two Chief World Systems, Ptolemaic and Copernican, proposing indeterminately the philosophical and natural reasons for the one as well as for the other side" (Galilei 1632, title page).

The intended message here is obviously that the book is merely discussing all the arguments for and against the two views, without defending or criticizing either side. This message is reinforced many times by appropriate reminders interspersed in the body of the work (for a comprehensive review, see Finocchiaro 1980: 12–18). The point of this rhetoric is clear. Galileo wanted to make sure he was not seen as defending Copernicanism, and thus violating the anti-Copernican decree of 1616.

The reason why this rhetoric did not work—why it was seen as "mere" rhetoric—is that in the book's actual content the pro-geostatic arguments are criticized and the pro-geokinetic ones favorably evaluated, and to do this is to defend the geokinetic theory. That is, appearances to the contrary, the book's content amounts to a defense of Copernicanism. The ineffectiveness of this rhetoric was one of the many factors that led to the Inquisition trial and condemnation of 1633.

Another rhetorical impression is conveyed by the book's Preface, together with a number of related passages, such as the passages on Scripture (in IIIB2) and on divine omnipotence (in IVA). The Preface tries to give the appearance that the book is a work of religious apologetics, aiming to show to the whole world that Catholics know all the scientific evidence about the subject; that they are aware that the pro-Copernican arguments are stronger; but that they regard these arguments as proving only that Copernicanism is more useful mathematically (not more likely to be true) than the geostatic view; and therefore that the 1616 decree was the result not of scientific ignorance but of religious motivations. One of these was the awareness of divine omnipotence, and accordingly the book's ending states the favorite objection of Pope Urban VIII.

However, such rhetoric of religious apologetics was no more effective than the rhetoric of indecision in the book's title. The difficulty was that, although the scientific and philosophical arguments admittedly did not establish with certainty the physical truth of the geokinetic theory, they did show not only that it was more mathematically convenient, but that it was also more likely to be physically true than the geostatic view; that is, Galileo was defending not merely the mathematical usefulness but also the physical truth of Copernicanism. It followed that, although the 1616 decree may not have been the result of ignorance, Galileo was both violating it in fact and implicitly criticizing it as erroneous.[4]

A third rhetorical impression derives from a few scattered remarks Galileo occasionally makes, to the effect that what he is doing with regard to the earth's motion is to give a strict demonstration or conclusive proof, and that this is what is generally required in scientific inquiry. For example, at the beginning of the presentation of Galileo's explanation of the tides, Salviati declares:

> In regard to the manner in which these effects should follow as a consequence of the motions which naturally belong to the earth, not only must they find no repugnance or hindrance, but they must follow easily; indeed, not only must they follow with ease, but with necessity, so that it is impossible for them to happen otherwise; for such is the character or mark of true natural phenomena. We have established the impossibility of explaining the motions we see in the water while

simultaneously maintaining the immobility of the containing vessel; so, let us go on to see whether the motion of the container can produce the effect and make it happen in the way it is observed to happen.

(FIN 288; cf. DML 492)

Analogous remarks are expressed at the beginning and end of the discussion of the explanation of the annual cycle of sunspot paths (DML 400–401, 413–14).

This rhetoric derives partly from the Aristotelian ideal of science as demonstration, which Galileo inherited from the background of his cultural milieu (see Wallace 1984, 1992a, 1992b), but which he was in the process of revising. In part, the rhetoric of demonstration may be an attempt to evade the anti-Copernican decree, insofar as a rigorous proof of a conclusion (if it is really rigorous and really a proof) is not really a defense of the conclusion, but only an exhibition of mathematical and logical relationships.

Moreover, unlike the passage just quoted, such rhetoric is not always expressed by Galileo's main spokesman (Salviati). For example, at the end of the discussion of sunspot paths, it is Sagredo who evaluates the argument as demonstratively necessary; and, as discussed earlier (Chapter 9.1), in the methodological discussion of critical reasoning, it is Simplicio who expresses the demonstrativist ideal. Thus, when the dramatic aspects of the discussion are taken into account, the rhetorical appearances may not correspond to Galileo's own intentions. Even when the speaker is Salviati, the rhetorical appearance may not survive critical analysis; for example, in the passage last quoted the alleged necessity refers more likely to the relationship between the geokinetic hypothesis and the tidal effects, rather than to the geokinetic hypothesis per se.

Finally, the rhetoric of strict demonstration is to some extent a rhetorical exaggeration on Galileo's part, on those occasions when he was advancing an argument which he felt to be especially strong.

At any rate, such rhetoric of strict demonstration failed for the simple reason that Galileo's arguments, however ingenious and strong, do not in fact amount to a strict demonstration, and this fact was known to him and readily perceivable by others. His arguments for the earth's motion are not as flimsy as his contemporary enemies and modern critics claim,[5] but they are certainly not demonstrative proofs. One important rhetorical incoherence is that, if any of his arguments for the earth's motion were really conclusive (even in his own mind), he would not be giving so many of them.

10.3 RHETORIC OF ANTI-RHETORIC

Let us now consider another especially relevant example of this type of rhetoric, which I am calling image projection; these considerations will pave the way for the examination of the other two types. The example is the rhetoric of anti-rhetoric already quoted above for the purpose of introducing the distinction among these three kinds. We have already seen that in that initial quotation there is rhetoric critical of the arts of eloquent expression and persuasive argumentation. The purpose now is to determine the effectiveness of this rhetoric of anti-rhetoric.

One can first analyze more deeply the internal coherence of the impression projected by that initial passage, to determine whether any message does get through after all. One difficulty is to ask what is being contrasted to Demosthenes's eloquent expression and Aristotle's persuasive argumentation. Presumably it is necessary demonstration. But, if so, it cannot be denied that "intellectual subtlety" is often required to grasp it, and so the "average intellect" is unlikely to be "fortunate" enough for the task.

It is also useful to examine how the book's readers perceived such rhetoric, even though their perceptions need not be accepted uncritically. The fact is that many were not impressed. In particular, the response of one reader is especially relevant because it contains a good argument why some rhetoric is sometimes needed in scientific inquiry. The respondent was Antonio Rocco who, in 1633, published a book against Galileo's *Dialogue* entitled *Philosophical Exercises* (reprinted in FAV 569–750).

One can also examine how the message in that initial, Galilean passage relates to others. For example, since the passage is contrasting the rhetoric being rejected to the idea of necessary demonstration, it could be taken as part of the book's rhetoric of strict demonstration, previously mentioned. If carried out sufficiently extensively, such comparative analysis would bring us close to identifying one element of the book's actual content with regard to the methodology of scientific inquiry. The result would be, as we saw earlier (Chapter 9.4), that the *Dialogue* advocates primarily the importance of critical reasoning and the judicious use of mathematics, rather than demonstrative proof and the mathematicist conflation of the world of numbers with the world of nature.

The anti-rhetorical passage initially quoted is not the only example of an attempt to project an anti-rhetorical image. There are two others. One of these occurs at the very beginning of the book (IA1). There, Galileo criticizes Aristotle's argument that the number three has a special perfection about it because three is the number of parts which every complete thing has (namely, beginning, middle, and end); because three is the number used in making sacrifices to the gods; and because three is the minimum number of things required before the word "all" can be used to refer to them collectively. Salviati reacts to this by saying:

> To be frank, in all these considerations I do not feel compelled to grant anything other than that whatever has a beginning, a middle, and an end may and should be called perfect; but I feel no inclination to grant that, because beginning, middle, and end are three in number, therefore the number three is a perfect number and has the faculty of conferring perfection to whatever has it. For example, I do not understand how or believe that in regard to number of legs three is more perfect than four or two; nor is it the case that the number four is an imperfection for the elements, and that it would be a greater perfection if they were three in number. Therefore, it would have been better to leave these subtleties to the rhetoricians and prove his conclusions with a necessary demonstration, for this is the appropriate thing to do in the demonstrative sciences.
>
> (FAV 35; cf. DML 11)

Here Galileo seems to view rhetoricians as traffickers in silly arguments, like the Aristotelian ones about the perfection of the number three, and as having no serious business to transact in "demonstrative" science.

Notice, however, that he expresses this anti-rhetorical sentiment at the end of a brilliant piece of rhetoric on his own part, namely a very convincing refutation of the Aristotelian claim about the perfection of the number three. He is therefore practicing rhetoric in the sense of persuasive argumentation, and what he is objecting to is the fact that the Aristotelian argument is unpersuasive.

It could be said, therefore, that Galileo is not only not practicing what he preaches, but also practicing the opposite of what he preaches. Then we would get into the problem of whether we should derive a lesson from his words or his deeds. Because his rhetorical practice is so prevalent and his anti-rhetorical pronouncements so few, the choice would be easy.

However, there may be a better way of analyzing the situation. Galileo's anti-rhetorical pronouncements are inconsistent with his rhetorical practice only if those pronouncements are unduly universalized into the rule that rhetorical persuasion is never appropriate in science. But perhaps all he is saying is that rhetoric is of secondary importance, which is consistent with its being occasionally appropriate. One of these occasions might be when another person has attempted some rhetorical persuasion that turns out to be incorrect for some reason. Then, like Galileo replying to Aristotle's argument about the number three, one may have to practice rhetoric to deal with rhetoric.

Thus, the lesson emerging from the Galilean passage just discussed is that rhetoric in the sense of persuasive argumentation has some role to play in science. One of these roles is to answer rhetoric with rhetoric. But this is not the only role. Some time ago, Thomas Kuhn (1922–96) argued that persuasion is the essential activity when a scientist is faced with paradigm choice; that is, with having to choose between one world view and a fundamentally different one.[6] This was exactly Galileo's predicament, and so indeed we find the whole *Dialogue* to be essentially a long piece of persuasive argumentation designed to justify the superiority of the geokinetic over the geostatic world view.

10.4 ELOQUENT EXPRESSION

A similar tension between anti-rhetorical words and rhetorical deeds exists within what seems to be the only other explicitly anti-rhetorical passage in Galileo's *Dialogue*. However, the passage is even more valuable as an illustration of eloquent expression. The context is as follows.

As we have seen, one of the issues in the Copernican controversy was whether there are essential differences between heavenly and terrestrial bodies. According to Aristotle's earth–heaven dichotomy, besides differences in natural motion, elementary composition, weight, luminosity, and susceptibility of physical changes, heavenly bodies had a purity, perfection, and nobility which terrestrial bodies did not possess.

Galileo criticized the earth–heaven dichotomy in several ways. One of his criticisms (IA6) was that the doctrine of heavenly unchangeability is false, as the telescopic evidence of sunspots and lunar mountains shows; this was primarily a type of scientific and observational criticism. Another Galilean criticism (IA5) pointed out that the doctrine of the earth–heaven dichotomy was self-contradictory; for it relied on the theory of change deriving from contrariety, and this theory implied that the contrariety between changeable terrestrial bodies and unchangeable heavenly bodies should produce changes for all of them. This was a philosophical and logical criticism.

Besides these observational and logical objections, Galileo also advances the following rhetorical criticism:

> Further, of the emptiness of such rhetorical conclusions, we have spoken many times. Is there anything more foolish than saying that the earth and terrestrial elements are relegated and separated from the heavenly spheres, but confined inside the lunar orb? Is not the lunar orb a heavenly sphere and, as they themselves agree, located in the middle of all the others? What a way of separating the pure from the impure and the sick from the healthy—to give those who are infected room at the heart of the city! And I thought that the lazaretto should be located as far as possible! Copernicus admires the arrangement of the parts of the universe because God placed the great lamp, which was to give the most light everywhere in his temple, at its center and not on one side ... But, please, let us not confuse these rhetorical flowers with solid demonstrations, and let us leave them to the orators, or rather to the poets, who with their

pleasantries know how to praise highly things that are very vile and even pernicious.

(FAV 292–93; cf. DML 311–12)

This passage begins and ends with the type of anti-rhetorical expressions which we have called the rhetoric of anti-rhetoric. However, for the most part it elaborates a type of rhetorical criticism of the geocentric position. Galileo is making fun of the idea of separating the pure from the impure by placing the impure in the middle, surrounded by the pure. I find the criticism extremely effective, and categorize it as involving primarily eloquent expression.

Further, I see nothing methodologically objectionable with employing this type of rhetoric, in the way Galileo does; that is, to use it in addition to empirical considerations, critical reasoning, and persuasive argumentation. Of course, it may be employed improperly, if, for example, it were used as a substitute for argumentation, reasoning, and observation. However, then the problem would be not with eloquent expression per se, but with some impropriety stemming from elsewhere.

Let us consider another example of eloquent expression. As we have seen, at the time of Galileo, the tides continued to puzzle natural philosophers, and no completely adequate explanation was available. In the Fourth Day, Galileo elaborates a theory that explains them in terms of the earth's motion, thus also providing what he felt to be one of his best arguments in favor of Copernicanism. The discussion begins by briefly criticizing alternative explanations (IVA). One of these is in terms of lunar heat increasing the temperature of sea water, and causing it to expand and thus to rise. One of Galileo's criticisms of this is the empirical one of inviting anyone to test the temperature of water at high and at low tides, and see that there is no difference. To this he adds, referring to proponents of this theory, the following rhetorical gem: "tell them to start a fire under a boiler full of water and keep their right hand in it until the water rises by a single inch due to the heat, and then to take it out and write about the swelling of the sea" (FIN 284; cf. DML 488).

The role of eloquent expression is not solely destructive. It is even more useful for constructive purposes. In the *Dialogue*, the best examples of constructive eloquence are the statements of the following arguments: the geokinetic argument from the negative correlation between luminosity and mobility (IIC5); the basic argument in favor of the heliocentrism of planetary revolutions (IIIC1); the explanation of the seasons (in IIIB3, specifically at DML 452–61); and the explanation of the monthly and annual periods of the tides (IVD). However, in the nature of the case, these cannot be summarized without destroying the eloquence of expression, or without recreating another original instance of eloquent expression.

As a final example of eloquence, it is instructive to examine a passage where the purpose of the eloquent expression is different from those considered so far. In the discussion of heliocentrism (IIIC1), at one point Galileo hurls an insult at some opponents of Copernicanism by calling them "men whose definition contains only the genus but lacks the difference" (FIN 233; cf. DML 380).[7] This is not much more than name calling, though it is very clever. Its meaning is as follows.

In traditional logic, definitions were given by identifying the genus and the species (or specific difference) to which the thing to be defined belongs; the genus is a broader category of classification, and the specific difference is a subdivision within the genus. Even the branch of modern biology called taxonomy still follows this procedure to some extent; for example, it defines man as *homo sapiens*, namely as belonging to the genus *homo* and the species *sapiens*. Traditional Aristotelian doctrine defined man as "rational animal," namely as belonging to the genus "animal" and the species "rational." Now, if from this definition of man the species is removed, we are left with "animal"; thus a person whose definition ("rational animal") contains only the genus ("animal"), but lacks the (specific) difference ("rational"), is an alleged rational animal who is not really rational but only a mere animal. In short, Galileo is here engaged in name-calling, subtle and unprosaic to be sure, but name-calling nonetheless; he is calling some of his opponents simply animals (or perhaps irrational animals).

This sort of rhetoric probably has no place in scientific inquiry, although it may be unavoidable for individual scientists to engage in it occasionally, while in the heat of a dispute. However, such rhetoric makes the *Dialogue* a more interesting book, and gives it a value in the aesthetic dimension, making it a work of art susceptible of being appreciated as literature.

10.5 RHETORICAL COMMUNICATION VS. METHODOLOGICAL REFLECTION

So far in this chapter, I have analyzed an aspect of Galileo's *Dialogue* which I had previously ignored. This has required defining, illustrating, and to some extent interrelating three types of rhetoric: the communication of impressions, persuasive argumentation, and eloquent expression. However, the analysis would be incomplete without some examination of how these rhetorical arts relate to the book's other aspects previously analyzed.

Let us consider, once again, Galileo's assertion that "in the natural sciences the art of oratory is ineffective" (DML 61). This may be rephrased as the claim that rhetoric is ineffective in natural science, and would imply that one should not use rhetoric in natural science. This assertion was regarded as part of his attempt to project an anti-rhetorical image—of his rhetoric of anti-rhetoric.

Consider also the assertion which I quoted as part of the evidence for Galileo's rhetoric of strict demonstration: " ... that it is impossible for them to happen otherwise ... such is the character or mark of true natural phenomena" (FIN 288; cf. DML 492). This could be rephrased by saying that the mark of true natural phenomena is their necessary truth, which would imply that the aim of physical science is to provide strict demonstrations of natural phenomena.

I have already suggested that these two assertions are different sides of the same coin because strict demonstration seems to be what Galileo opposes to the rhetoric which he claims to be rejecting. The point to stress next is that these assertions are instances of methodological reflection, for they formulate principles about the nature of scientific knowledge, as well as general methodological principles which one may follow in the search for truth.

In this regard, they are like the many methodological remarks which fill the *Dialogue*, and which we analyzed earlier (Chapter 9).

For example, in the discussion of rationality and critical reasoning, Galileo's book contains the assertion that there is a one-to-one correspondence between physical truth and good reasoning on the one hand and physical falsity and bad reasoning on the other (DML 151; cf. Chapter 9.1 above). In the discussion of authority and independent-mindedness (cf. Chapter 9.2), we find the remark that there is nothing "more shameful in a public discussion dealing with demonstrable conclusions than to see someone slyly appear with a textual passage (often written for some different purpose) and use it to shut the mouth of an opponent" (FIN 127; cf. DML 131). In the discussion of sensory observation and theoretical reason, there is the claim that it is fitting that sensory experience should have priority over intellectual theorizing (DML 57; cf. Chapter 9.3 above). And in the discussion of mathematical truths and physical reality (cf. Chapter 9.4), Galileo asserts that "to want to treat physical questions without geometry is to attempt to do the impossible" (FIN 193; cf. DML 236).

Now, we have also seen that such assertions could be interpreted to attribute to Galileo a methodology which he does not really espouse. The first one of these four claims could be taken as an expression of a simple-minded rationalism unappreciative of the importance of critical reasoning. The second could be interpreted to imply a complete rejection of all authority. The third could be construed as a commitment to a naïve empiricism. And the fourth one could be understood as a version of extreme mathematicism.

However, I argued that such attributions would involve misinterpretations of Galileo. The simple-minded correspondence between physical truth and human reasoning is asserted by Simplicio, but is later appropriately qualified by Sagredo, and it contradicts Galileo's whole procedure in his book. The complete rejection of authority would involve taking the above quoted remark out of context, divorcing it both from other remarks about authority and other things Galileo does in his book. The naïve empiricism would reflect only one side of the methodological situation and unduly neglect Galileo's inclinations toward intellectual theorizing. And the mathematicist ideal would be an exaggeration of the quoted sentence and would be unable to do justice to the qualitative and nonmathematical side of Galileo's procedures.

Whether or not my own interpretation and my criticism of alternative construals are correct, the important point is that the understanding and appreciation of the methodological aspect of the *Dialogue* is not easy and generates controversy among readers. Earlier, in the discussion of the book's methodological reflections, the alternative methodological interpretations were rejected as being either one-sided, or exaggerated out of proportion, or taken out of context, or failing to correspond to Galilean practice as a whole. There was then no talk of rhetoric, as there has been in this chapter for the case of Galileo's rhetoric of anti-rhetoric and his rhetoric of strict demonstration. Why not? Is there a difference between the earlier cases and these?

To answer these questions, let us note that we could apply to the earlier cases the notion of rhetoric as the projection of impressions. We could say that there is also in Galileo's book a rhetoric of simple-minded rationalism, a rhetoric of total rejection of authority, a rhetoric of naïve empiricism, and a rhetoric of mathematicism. Using such rhetorical terminology, the key point of my earlier analysis was that all these rhetorics do not correspond to reality, namely the reality of Galileo's methodological practice and the totality of his methodological reflections. For, besides a rhetoric of naïve rationalism, there is also a rhetoric and a practice of judicious rational-mindedness and critical reasoning; besides a rhetoric of total anti-authoritarianism, there is a rhetoric and a practice of using authorities as sources of arguments and methodological principles; besides a rhetoric of empiricism, there is a rhetoric of apriorism and a practice of judiciously combining empirical and intellectual procedures; and besides a rhetoric of mathematicism, there is a rhetoric of the pitfalls of applying mathematical truths to physical reality and a practice of a limited use of the mathematical approach.

Similarly, the rhetoric of anti-rhetoric was criticized partly as internally incoherent and partly as inconsistent with Galileo's overall practice. And the rhetoric of strict demonstration was criticized as also contradicted by his general practice; it could also be criticized as inconsistent with his rhetoric of epistemological modesty (cf. IB4) and with the rhetoric of probabilism which we find in the book's Preface and in the discussion of the simplicity arguments for terrestrial rotation (in IIA).

Thus, there is no essential difference between the cases of methodological reflections examined earlier (Chapter 9) and the rhetoric of anti-rhetoric and of strict demonstration examined above (in this chapter). The main difference lies in my evaluation of the corresponding methodological principles and of the corresponding methodological interpretations of Galileo. In my earlier discussion of methodological reflections, I was focusing on passages and issues where we could derive useful methodological lessons from our reading of Galileo, and in the process I was criticizing various one-sided readings which were based on corresponding rhetoric on his part. In the present discussion of rhetoric, I have been focusing on examples of methodological reflections which do not lend themselves to deriving useful lessons, in part because the corresponding Galilean pronouncements largely contradict his own practice and so should be regarded as primarily rhetoric (in the sense of attempts at public relations).

In fact, there is an overlap between rhetoric in the sense of image projection and methodological reflection. Some rhetoric (when the subject matter is methodological procedures) is simply part of methodological reflection. One could expand the definition of methodological reflection to include the communication of methodological principles, besides their formulation, analysis, evaluation, and application. The communication of methodological principles (namely, the projection of methodological images) is not always effective. In some cases, for some reason or other, the methodological assertions fail to get really projected. The reasons may involve considerations of context, overall balance of textual evidence, correspondence between words and deeds, and other considerations involving one-sidedness and exaggerations. When rhetoric is ineffective, one may speak of "mere" rhetoric. To use such a locution is to express a negative evaluation of the attempt at communication being considered. However, not all rhetoric is mere rhetoric.

10.6 PERSUASIVE ARGUMENTATION VS. CRITICAL REASONING

As just discussed, rhetoric in the sense of image projection raises the obvious question of its connection with methodological reflection, given that the projection of methodological and epistemological images is an obvious area of overlap. Analogously, rhetoric in the sense of persuasive argumentation raises the obvious question of its relationship to critical reasoning, since it is even more obvious that argumentation provides an analogous area of overlap.

Let us recall that reasoning is the mental process of interrelating thoughts in such a way that some are based on or follow from others; that argument or argumentation is the special case of reasoning when conclusions are supported by reasons and/or defended from objections; and that critical reasoning is the special case of reasoning when arguments are interpreted, analyzed, evaluated, or self-reflectively presented. Now, persuasive argumentation sounds and is the special case of argument when it has the property of persuasiveness or persuasive force.

Persuasiveness is, first of all, a positive evaluative property, as distinct from a structural property. That is, to say that an argument is persuasive is to appraise it as good in some respect; it is not to give an interpretation of what its parts are and how they interrelate, so as to provide an understanding of what the argument is. Second, persuasiveness is a matter of degree, rather than an all-or-none affair; a gradual, not a discrete, notion; that is, arguments are capable of being more or less persuasive.

The persuasive force of an argument refers to the extent to which the audience comes to accept the conclusion based on the supporting reasons and/or criticism of the objections advanced in the argument. The extent of such audience acceptance depends on at least two factors: the number of persons who did not otherwise accept the conclusion but do so as a result of the argument, and the increase in the strength of the belief in the conclusion. That is, an argument is persuasive if and to the extent that it increases the acceptance of the conclusion, i.e., causes the conclusion to become accepted more widely and/or more strongly.

As just defined, persuasiveness refers to psychological and sociological phenomena about people's beliefs and, in that sense, to a factual and empirical situation. However, these facts involve mostly what goes on in people's minds, and so they are not open to direct observation. Therefore, one will often resort to indirect methods of ascertaining how much more widely or strongly a conclusion is believed. One of these indirect methods involves exploring the extent to which the audience *should* come to believe the conclusion as a result of the argument. The assumption here is that people are rational, in the sense that they actually believe what they should. This assumption would be questionable if it were construed as the universal generalization that each person always does this on all occasions. However, the assumption is plausible if taken to mean that most people most of the time do this; that they normally or typically do this.

Thus, the definition of persuasive force should be expanded: the persuasiveness of an argument refers to the extent to which the audience *comes or should come* to believe, or accept, the conclusion based on the supporting reasons and/or on criticism of the objections advanced in the argument. More generally, we might say that persuasiveness refers to the extent to which the audience does come, or should come, or would come, or is likely to come to believe the conclusion.

Because of this broader definition, and because people's mental activities are not directly observable, the assessment of persuasive force consists to some extent of argumentation and counter-argumentation about the argument being considered. How large this extent is would depend on one's orientation. Scholars in communication studies tend to focus on the empirical approach, whereas philosophers tend to focus on conceptual considerations. Such conceptual considerations need not be purely formal or apriorist. To do so would be an attempt to reduce persuasive force to logical force or formal validity. However, one key point about persuasiveness is that the increased acceptance of the conclusion, based on the supporting reasons and/or on the criticism of objections, may be due to logical factors, but may also be due to psychological, aesthetic, or other nonlogical factors. One of these nonlogical factors is eloquent expression.

Let us apply these ideas to the relevant rhetorical passages from Galileo's book introduced earlier. The passage on the perfection of the number three was discussed earlier as an illustration of both the rhetoric of anti-rhetoric and persuasive argumentation. The persuasive argumentation consisted of the criticism of the perfection of the number three. Logically, the argument may be reconstructed as follows: it is wrong to think that "the number three is a perfect number and has the faculty of conferring perfection to whatever has it" (FAV 35; cf. DML 11), because having three legs apparently does not make animals more perfect than having two or four legs and, similarly, the existence of four elements (in the terrestrial region), according to the geostatic world view, has never been thought by anyone to be a sign of imperfection for that viewpoint. The persuasive force of the argument is due to the fact that Galileo is refuting a generalization about the number three by citing two counterinstances of cases where the number three does not confer perfection. This refutation is a matter of logical force, and this logical force generates the argument's persuasiveness.

Another rhetorical example was Galileo's comparison of a geocentric universe to having a lazaretto in midtown. This passage was given as an example of both his rhetoric of anti-rhetoric and eloquent expression. The eloquence stems from the clever image of a lazaretto in midtown, and from the striking character of its comparison to the geostatic universe, where the impure earth is in the middle of the pure heavenly bodies. The eloquence of these images does increase one's tendency to think of the geostatic system as an inappropriate arrangement. To see that we have persuasive force due to eloquence, we need to reconstruct the passage as an argument from analogy: the geostatic universe is unlikely to be true because it is like having a lazaretto in midtown, and such city planning is very inappropriate. This argument seems to have a persuasive force above and beyond any which it might derive from logically considering the analogy; I say that the extra persuasiveness is due to the cleverness and eloquence of the analogy.

The third example of eloquent expression given above was the criticism of the heat theory of tides. The proponents of this theory were invited to test it by burning their hand in a kettle of boiling water, while waiting for the level of the water to rise in a tidal-like fashion due to the heat. We have already seen that there is a purely empirical component to Galileo's criticism, which he expressed in standard neutral manner, referring to the temperature of water during high and low tides. However, the clever invitation has a life of its own and adds something else to the criticism. That is, the passage could be reconstructed logically in terms of the following argument: it is false to account for the tides by saying that they are caused by heat because, first, seawater in high tides is no warmer than in low tides, and second, the expansion of water due to heat would be insufficient to make the level of seawater rise as much as it does during high tides; in turn, the latter fact can be ascertained by doing the kettle experiment. The persuasive force of this last subargument is largely due to the eloquence of the suggestion.

What these examples show is that rhetoric in the sense of eloquent expression is related to rhetoric in the sense of persuasive argumentation. They also show that persuasive argumentation is simply a special case of argumentation, and is in that sense related to critical reasoning.

NOTES

1 Descartes to Mersenne, 14 August 1634, in Galilei 1890–1909, vol. 16: 124–25; also in Descartes 1897–1913, vol. 1: 303–6.

2 Cf. Feyerabend 1988, Finocchiaro 1980, Hill 1984, Jardine 1991, Moss 1993, Moss and Wallace 2003, Vickers 1983.

3 These correspond to the three present-day disciplines of communication studies, English composition, and philosophical rhetoric in the sense of the "new rhetoric" (Perelman and Olbrechts-Tyteca 1969).

4 For more details on this religious rhetoric in the *Dialogue*, see Finocchiaro 1980: 6–12.

5 For a discussion of this issue, and references to the literature, see Finocchiaro 2010b: 235–43.

6 Kuhn 1970: 152. Cf. Brown 1977, Margolis 1987, Shapere 1984.

7 In this case, the point will be completely missed if one relies on Drake's translation (DML 380, DCA 327), as shown in Finocchiaro 1980: 244; the point is also missed in Shea-Davie 307; even Strauss, who gets the German translation right (Strauss 342 = Sexl-Meyenn 342), seems to miss the essential point in his comment (Strauss 551 n. 34 = Sexl-Meyenn 551 n. 34), as shown in Finocchiaro 1980: 230–31.

Part IV

CONCLUSION

11

HISTORICAL AFTERMATH AND ENDURING LEGACY

11.1 HISTORICAL REPERCUSSIONS

Although the 1633 condemnation of Galileo and prohibition of the *Dialogue* ended the original affair, it also started a new controversy continuing to our own day—about the facts, causes, issues, and implications of the original episode. This subsequent controversy partly reflects the original issues, such as the reality of the earth's motion and the astronomical authority of Scripture. But it has also acquired a life of its own, with debates over whether Galileo's condemnation was right; why he was condemned; whether science and religion are incompatible; whether individual freedom and institutional authority must always clash; etc. The original affair is the aspect of the Copernican revolution consisting of Galileo's contributions to it. The subsequent affair is much more complex because of the longer historical span, the broader inter-disciplinary relevance, the greater international and multi-linguistic involvement, and the ongoing cultural import. Simplifying, that historical aftermath may be highlighted as follows (for details, see Finocchiaro 2005b, 2010b: 155–228).

One strand of that story involves the key scientific claim which is contained in the *Dialogue* and for which Galileo was condemned, namely the proposition that the earth moves. The condemnation ignited a scientific controversy, which had existed since Copernicus, but which now took a more definite and intense form—more definite because it now focused on whether the earth really moves and whether this motion can be proved observationally by terrestrial or astronomical evidence, and more intense because scores of books were published, new experiments devised, new arguments invented, and old arguments re-hashed.

In 1687, Newton brought the Copernican revolution to a climax with a synthesis of the work of Copernicus, Kepler, Galileo, Descartes, Huygens, and others, in a book entitled *Mathematical Principles of Natural Philosophy*. The Newtonian system has two important geokinetic consequences. First, the relative motion between the earth and the sun corresponds to the actual motion of both bodies around their common center of mass; but the relative masses of the sun and the earth are such that the center of mass of this two-body system is a point inside the sun; so, although both bodies are moving around that point, the earth is circling the body of the sun. Second, the daily axial rotation of the earth has the centrifugal effect that terrestrial bodies weigh less at lower latitudes and least at the equator, and the whole earth is bulged at the equator and flattened at the poles; these consequences were verified by observation.

However, the controversy over the earth's motion did not end then because the Newtonian proofs were indirect and theoretical. The search for direct experimental evidence of the earth's motion continued. This led to the discovery of the aberration of starlight by James Bradley in 1729, showing that the earth has translational motion in space; the discovery, by Giambattista Guglielmini in 1789–92, that freely falling bodies are deflected eastward away from the vertical by a small amount, verifying terrestrial axial rotation; the discovery of annual stellar parallax by Friedrich Bessel in 1838, proving the earth's revolution in a closed orbit; and the invention of Foucault's pendulum in 1851, providing a spectacular demonstration of terrestrial rotation.

Another strand of the subsequent affair involves actions by the Catholic Church designed to repeal the censures against the Copernican doctrine and books. In 1744, Galileo's *Dialogue* was republished for the first time with ecclesiastic approval, as the fourth volume of his collected works; the text was preceded by the Inquisition's sentence and Galileo's abjuration of 1633, by an apologetic editorial preface, and by an erudite hermeneutical introduction written by a contemporary biblical scholar. In 1757, with the approval of Pope Benedict XIV, the *Index of Prohibited Books* dropped from the list of general prohibitions the clause "all books teaching the earth's motion and the sun's immobility," although it continued to include several previously prohibited books, including Copernicus's *Revolutions* and Galileo's *Dialogue*. In 1820, the Inquisition gave the imprimatur to an astronomy textbook by a professor at the University of Rome that presented the earth's motion as a fact, thus overruling the objections of the chief censor in Rome. In 1822, the Inquisition ruled that in the future this official must not refuse the imprimatur to publications teaching the earth's motion in accordance with modern astronomy. In 1833, while deliberating on a new edition of the *Index*, Pope Gregory XVI decided that it would omit Copernicus's *Revolutions* and Galileo's *Dialogue*; thus, the 1835 edition of the *Index* no longer listed them. This was the final and complete retraction of the book censorship begun in 1616 and expanded in 1633.

However, besides the substantive scientific issue of the earth's motion, the original Galileo affair also embodied a question of principle, namely whether Scripture is an astronomical authority as well as being one for matters of faith and morals. This question is partly philosophical and partly theological. It too culminated in 1633, when the Inquisition's sentence convicted Galileo in part because the *Dialogue* implicitly denied the scientific authority of Scripture. Thus, one strand of the subsequent controversy involves this principle.

One crucial episode in this strand is that eventually the Church ended up agreeing with Galileo. In 1893, in the encyclical *Providentissimus Deus*, Pope Leo XIII advanced a view of the relationship between biblical interpretation and scientific investigation that corresponds to the one elaborated in Galileo's *Letter to the Grand Duchess Christina.* Although Galileo was not even mentioned in the encyclical, the correspondence was easy to detect for anyone acquainted with both documents; so the encyclical was an *implicit* vindication of Galileo's principle that Scripture is not a scientific authority.

A century later, the vindication was made *explicit* in Pope John Paul II's rehabilitation of Galileo in 1979–92. Although this rehabilitation was incomplete, informal, and problematic in several ways, on the hermeneutical issue John Paul was clear. In a 1979 speech, he declared that "Galileo formulated important norms of an epistemological character, which are indispensable to reconcile Holy Scripture and science" (John Paul II 1979: 10). And in a 1992 speech, the pope specified: "the new science, with its methods and the freedom of research that they implied, obliged theologians to examine their own criteria of scriptural interpretation. Most of them did not know how to do so. Paradoxically, Galileo, a sincere believer, showed himself to be more perceptive in this regard than the theologians who opposed him" (John Paul II 1992: 2).

This historical strand also includes the response by various Protestant denominations to the Galilean principle that Scripture is not a scientific authority, especially the question of the applicability of this principle to the theory of evolution in biology. Clearly there are some similarities and some differences between the status of the geokinetic theory in Galileo's time and for some time thereafter and the status of evolutionary theory since Charles Darwin's epoch-making contributions, and between the limitation of biblical authority in celestial mechanics and in the life sciences. Here, it is intriguing that the Catholic Church seems to have displayed a relatively more enlightened attitude toward evolutionary theory than evangelical Protestant fundamentalists, and some have attributed this to the Church's involvement in, and learning from, the Galileo affair.

This brings us to one final strand of the historical aftermath: the condemnation of Galileo as a person, as distinct from the prohibition of the book. This strand consists of various ecclesiastic attempts to revise the trial or rehabilitate him. It is the most elusive, complex, and controversial aspect of the story; and it is far from being closed.

This story began immediately after Galileo's death, when questions were raised about whether a convicted heretic like him had the canonical right to have his last will and testament executed, and whether he could be buried on consecrated ground. These issues were decided in his favor. But another question was not, namely whether it was proper to build an honorific mausoleum for him in the church of Santa Croce in Florence, which the Tuscan government was considering. This was vetoed by the Church when Galileo died in 1642. However, it finally happened in 1737.

Two centuries later, in 1942, the tricentennial of Galileo's death occasioned a first partial rehabilitation. In the period 1941–46, this was done by several clergymen who held the top positions at the Pontifical Academy of Sciences, Catholic University of Milan, Pontifical Lateran University in Rome, and Vatican Radio. They published accounts of Galileo as a Catholic hero who upheld the harmony between science and religion; who had the courage to advocate the truth in astronomy, even against the religious authorities of his time; and who had the religious piety to retract his views outwardly when the 1633 trial proceedings required his obedience.

In 1979, Pope John Paul II began a further informal rehabilitation that was not concluded until 1992. In two speeches to the Pontifical Academy of Sciences, and other statements and actions, the pope admitted that Galileo's trial was not merely an error but also an injustice; that, as already mentioned, Galileo was theologically right about scriptural interpretation, as against his ecclesiastical opponents; that pastorally speaking, his desire to disseminate scientific novelties was as reasonable as his opponents' inclination to resist them; and that he provides an instructive example of the harmony between science and religion. This rehabilitation was informal because the pope was merely expressing his personal opinions and not speaking *ex cathedra*. Moreover, it was partial because he deliberately avoided action regarding a formal judicial revision of the 1633 sentence. Finally, the rehabilitation was opposed by various elements within the Church, including some in the Vatican Commission on Galileo, which he had appointed in 1981, and which attempted to repeat many traditional apologias.

11.2 FROM COPERNICUS TO DARWIN, FREUD, AND BEYOND

The most widely-drawn lesson from the Copernican revolution in general, and from Galileo's *Dialogue* in particular, is the realization that mankind is not the physical center of creation, but rather inhabits an ordinary planet circling an ordinary star in an ordinary galaxy. This thesis was formulated with classic incisiveness by Sigmund Freud (1922: 240–41), who paired the Copernican lesson with an implication of Darwin's evolutionary theory: the demotion of the human species from the special place it had had in the phenomenon of organic life. Freud also speculated that his own discovery of the unconscious amounted to a comparable revolution in the domain of mental phenomena.

This thesis has recently been criticized, being called a myth, primarily for being anachronistic (Danielson 2009). The critic argues that at the time of the Copernican controversy, the "relocation" of the earth from the center of the universe to the third heliocentric orbit was not regarded as a demotion of humanity from the center of the cosmos, but rather as a promotion or ennobling; the reason is that the center was regarded as the place where all the waste of the universe ended up, and so it was better to be located away from the center. Allegedly, the Freudian thesis is a reinterpretation of the seventeenth-century transition from geocentricism to heliocentrism, formulated from the point of view of twentieth-century modes of thought.

Such criticism is engaged in an equivocation. For in the sixteenth and seventeenth centuries, there was an ambivalent attitude toward the center and the heavens, such that the center was better than the heavens in some respects, but worse in others. On the one hand, Aristotelian physics required that the center be occupied by the heaviest or grossest material. On the other hand, the Copernicans tried to draw an analogy between the central sun and the position of an emperor, who could oversee the whole realm from the center without moving much. This ambivalence was brilliantly exploited by Galileo in his eloquent image that the geocentric universe was like a town with a lazaretto at the center, as we saw earlier (Chapter 10.4). The criticism is attributing to the historical agents only one side of their ambivalent attitude, and to the moderns the other side. Moreover, part of the change in the Copernican revolution was a change in the attitude toward the center: from viewing the center as the place of waste and thus a bad place, to viewing it as a good, privileged position from where a ruler on the throne can oversee his domains. In short, the criticism is itself criticizable, since it is based on a one-sided analysis of the historical situation, which contained a mixture of attitudes, one of which did indeed correspond to the Freudian interpretation.

However, here my main point is that this lesson about the place of humanity in creation remains an open question deserving further reflection, especially with regard to the religious implications. And the lesson is open both in the sense that it is instructive to discuss what the Copernican revolution implies about the cosmological status of human beings, and in the sense that the Copernican lesson is liable to be expanded to other domains. That is, even if we accept Freud's interpretation of his own discovery of the unconscious and its relationship to the Copernican and Darwinian revolutions, his accomplishment is unlikely to be the end of this type of cultural lesson; for the current computer revolution may embody the next sobering lesson for humanity.

11.3 SCIENCE VS. RELIGION?

On another topic, a second enduring lesson occasioned by the *Dialogue* involves the question of what, if anything, its banning, and Galileo's trial in general, show regarding the relationship between science and religion. As traditionally interpreted, the affair epitomizes the conflict between science and religion. This interpretation is well known, but it is important to stress here that it has been advanced not only by relatively injudicious writers who have recently been widely discredited, e.g., John William Draper (1875) and Andrew Dickson White (1896), but also by such cultural icons as Albert Einstein (1953: 7), Bertrand Russell (1997: 31–43), and Karl Popper (1963: 97–98). At the opposite extreme, there is the revisionist thesis that the affair really shows the *harmony* between science and religion. This harmonious interpretation does not merely deny the traditional thesis but *reverses* it. Its most significant advocate is Pope John Paul II, for whom this was the key point he wanted to make in his rehabilitation of Galileo in 1979–92.

The harmony interpretation begins by distinguishing between the Catholic religion, as such, on the one hand, and men and institutions of the Church on the other. It then goes on to say that the injustices and errors were committed by men and institutions for which they and not the Church are responsible; so the conflict was between a scientist and some churchmen. With regard to the relationship between science and religion, the correct view is presumably the one elaborated by Galileo himself, which the Church later adopted as its own. That view says that God revealed himself to humanity in two ways, through His work and through His word. His word, namely Holy Scripture, aims to give us information which we cannot discover by examining His work. But to learn about His work, we need to observe it using our bodily senses and to reason about it with that other aspect of the Divine Work which is our mind. In short, Scripture is only an authority on questions of faith and morals, not on scientific factual questions about physical reality. In Galileo's trial, a key difficulty was the misunderstanding of these principles by the churchmen in power; once these principles are clarified, as Galileo himself ironically contributed to doing, the conflict between science and religion evaporates and continues to subsist only in the imagination of people who do not know better.

In contrast to both the conflict and harmony theses, I would claim that the trial did have *both* conflictual and harmonious aspects when viewed in terms of science and religion, but that these are elements of its *surface structure*, and that its most profound *deep structure* lies rather in the clash between *cultural conservation and innovation*. My argument is the following.

First, as already mentioned, the 1633 Inquisition sentence condemned Galileo for two beliefs: that the earth moves and that Scripture is not a scientific authority. The second issue involved a disagreement between those (like Galileo) who held and those (like the Inquisitors) who denied that it is proper to defend the truth of a physical theory contrary to Scripture. That is, if in this controversy we take the Copernican theory to represent science and Scripture to represent religion, then Galileo was the one claiming that there is no real incompatibility between the two, whereas the Inquisition was the one claiming that the apparent conflict was real. It follows that there is an *irreducible* conflictual element in Galileo's trial, between those who believed and those who denied that there is a conflict between Scripture and science. The irony of the situation is that it was the victim who held the more fundamentally correct view. However, insofar as that Galilean non-conflictual view is the more nearly correct one, then the content of that view suggests an important harmonious element in the affair.

Furthermore, both conflict and harmony exist at the level of the surface structure of the situation. If we move to a deeper cultural aspect, then we must point out that Galileo was not the only one who held there was no conflict, and that many of those who agreed with him were themselves churchmen. For example, the author of the first published (1622) defense of Galileo was Dominican friar Tommaso Campanella; and the author explicitly condemned in the 1616 Index decree was Carmelite friar Paolo Foscarini, whose book argued that the earth's motion is compatible with Scripture. That is, in Galileo's time, there was a division within Catholicism between those who did and those who did not accept the scientific authority of Scripture. A similar split existed in scientific circles. A further division existed in both domains with regard to the other main issue of Galileo's trial—the proposition of the earth's motion. Thus, rather than having an ecclesiastic monolith on one side clashing with a scientific monolith on the other, the real conflict was between two attitudes, criss-crossing both. The most fruitful way of conceiving the two factions is to describe them as conservatives or traditionalists on one side and progressives or innovators on the other. The real conflict was between these two groups. In this sense, Galileo's trial illustrates the clash between cultural conservation and innovation and is an episode where the conservatives happened to win. This conflict is one that operates in such other domains of human society as politics, art, economy, and technology. It cannot be eliminated without stopping social development; it is a moving force of human history.

Now, after Galileo's condemnation, as mentioned earlier, the interpretation and evaluation of the trial became a cause célèbre in its own right. Even those who nowadays advocate the harmony thesis (about the *original* episode), do not deny that the key feature of the *subsequent* affair was indeed a conflict between science and religion. For example, Pope John Paul II, believing that the lesson from Galileo's trial is the harmony between science and religion, wanted to stress this lesson in order to put an end to the subsequent, very real, but presumably unjustified, science versus religion conflict. Regarding this subsequent controversy, the science versus religion conflict is indeed an essential feature of it, much more of an integral part of it than of the original trial. However, underlying such surface structure there may be a cultural deep-structure; but in this case the deep structure is probably the phenomenon of the birth and evolution of cultural myths and their interaction with documented facts.

This nuanced account may disappoint those who seek simplicity and simple lessons. However, I believe that anything simpler or less nuanced is likely to be an oversimplification and hence misleading.

11.4 EINSTEIN AND SCIENTIFIC METHOD

Another enduring lesson derivable from the *Dialogue* involves using it as a methodological model in the search for truth and the acquisition of knowledge. This is possible both in the field of natural science and in other fields. Let us begin with an example from natural science.

The example comes from Einstein's remarks, which earlier (Chapter 8.1) I utilized to substantiate the book's wealth of scientific content. Now I want to elaborate a different aspect of his remarks. In fact, they can be reconstructed as the following argument: (1) In the *Dialogue*, Galileo rejected the hypothesis of the existence of a center of the universe as providing the explanation of the fall of heavy bodies. (2) In modern relativistic physics, the hypothesis of an inertial system as providing the explanation of the inertial behavior of matter is analogous to the one rejected in Galileo's *Dialogue*; for both hypotheses introduce a conceptual object which (a) does not have the same kind of reality that matter and fields do, and which (b) affects the behavior of real objects, without being affected by them. Therefore, (3) the hypothesis of an inertial system is as unscientific as the one about a center of the universe rejected by Galileo. Therefore, (4) the hypothesis of an inertial system should be rejected.

Here Einstein is referring to Galileo's critique of the geocentric argument from natural motion in the first part of the First Day of the *Dialogue*. His first premise is attributing to Galileo the criticism that this Aristotelian argument assumes that a center of the universe exists; although such a criticism is not explicit in the text, it is certainly plausible to make the attribution, and say that it is implicit. Then Einstein is assuming that Galileo is an appropriate scientific model, and goes on to detect an analogy between the situation he himself faces and the one faced by Galileo. Einstein's conclusion is that today he can do something analogous to what Galileo did then.

In this particular case, the analogy is not merely asserted, but also justified, by basing it on an appeal to a methodological principle under which both situations can be subsumed. In Einstein's words, this principle says that the introduction of an entity such that "it determines the behavior of real objects, but it is in no way affected by them ... is repugnant to the scientific instinct" (Einstein 1953: xiii). This principle, too, happens to have an analogue in Galileo. In fact, in a related passage, he explicitly applies the same principle to the case of the heavenly bodies (made of aether) that affect terrestrial bodies (made of earth, water, air, and fire) without being affected by them. With his inimitable mixture of scientific intuition, methodological acumen, rhetorical sensibility, and poetical imagination, Galileo remarks that "I do not see how the influence of the moon or sun in causing generations on the earth would differ from placing a marble statue beside a woman and expecting children from such a union" (DML 68–69).

Thus, Galileo's *Dialogue* can be used, and has been used, by working scientists as a model to follow in their own scientific investigations. Such modeling does *not* consist of a simple process of mechanically following easy recipes found in, or attributable to, that book; once again, such a process would amount to simple-mindedness and oversimplification, rather than true simplicity. Instead, the methodological utilization of the *Dialogue* involves the critical comparison and contrast of the scientific problem one is facing with some scientific problem faced by Galileo, and such a comparison-contrast is to be done in terms of both concrete substantive details and relevant methodological principles. This is a nuanced and judicious process, i.e., an exercise in judgment.

11.5 GALILEAN APPROACH TO THE GALILEO AFFAIR

Let us now illustrate the utilization of Galileo as a model, and of the *Dialogue* as a source of methodological lessons, in a field other than natural science. It happens to be a field which is doubly relevant to our concerns here, as I hope will become obvious.

To this end, let us recall the earlier analysis (Chapter 9) of the methodological content and significance of Galileo's *Dialogue*. I argued that Galileo preached and practiced a number of metho-dological principles, among which there were some which I labeled rational-mindedness, open-mindedness, fair-mindedness, judicious-mindedness, and critical reasoning. Rational-mindedness (or more simply, rationality, in one sense of this word) means the willingness and ability to *accept* the views supported by the most cogent arguments and the strongest evidence. Open-mindedness (or openness) is the willingness and ability to *know and understand* the arguments and evidence *against* one's own views. Fair-mindedness (or fairness, in one sense of this word) refers to being willing and able to *learn* from and appreciate the arguments and evidence against one's own views, even when one is attempting to refute them. Judiciousness (or judicious-mindedness) means being willing and able to be impartial and balanced, to avoid one-sidedness by properly taking into account all distinct aspects of an issue, and to avoid extremism by properly taking into account the two opposite sides of any one aspect. Finally, critical reason-ing is the skill of being willing and able to engage in reasoning aimed at the interpretation, evaluation, analysis, or self-reflective formulation of arguments.

I believe it is fruitful to follow this Galilean methodology in an important area of present-day concern. This area is the controversy about the 1633 trial and condemnation of Galileo and banning of the *Dialogue*; i.e., the cause célèbre that started then, continues to our own day, and tries to ascertain the facts, causes, consequences, issues, responsibilities, and lessons of the original episode. This is what on several previous occasions I have been calling the *subsequent* Galileo affair, by contrast with the *original* affair.

There is merit in this proposal because it seems undeniable that Galileo was successful in the context of the Copernican controversy, and so, if we can model our own approach to the subsequent controversy on the Galilean approach to the original controversy, we stand a good likelihood of success. Of course, this proposal does not deny that the two controversies are in some ways very different: the original one was in the fields of astronomy and physics, whereas the subsequent controversy is in the fields of history and philosophy. However, the point is to apply some of the formal and general features of the Galilean approach to a new and different situation.

The first step would be to follow rational-mindedness and focus on the arguments of both sides. In so doing, we also need to find some key claim which, as a result of such argumentation, is affirmed by one side and denied by the other. A promising candidate is the proposition that the Inquisition's condemnation of Galileo in 1633 was right. This is then the key issue of the subsequent controversy. Next, in accordance with the ideal of open-mindedness, one would focus on the subsequent arguments trying to justify his condemnation and defend the Church (i.e., the anti-Galilean arguments), to see whether they have any validity. The result would be the following sequence and framework of argumentation.

One initial response by critics of Galileo and pro-clerical apologists was to try to show that he had been scientifically wrong. For example, in 1642–48, a controversy developed regarding the correctness of his science of motion; a controversy that has been called "the Galilean *affaire* of the laws of motion … a second 'trial'" (Galluzzi 2000: 539; cf. Finocchiaro 2005b: 80–81). The result of this first "retrial" was a vindication of Galileo, who of course was dead by then, the controversy having ironically started the same year as his death (1642).

Similarly, in 1651, a Jesuit astronomer named Giovanni Battista Riccioli claimed that the Inquisition had been right and wise in condemning Galileo, both scientifically and theologically. Scientifically speaking, Riccioli argued that this was so chiefly because neither the Ptolemaic nor the Copernican, but rather the Tychonic, system was the correct one, and so Galileo was wrong in holding that the earth moves. Riccioli made a comprehensive examination of all the arguments to support his scientific choice. He even invented a new geostatic argument based on Galilean ideas—a Galilean argument against Galileo. In 1665, this argument engendered a controversy that lasted four years and spawned at least nine books.[1] Once again, the objections of Galileo's scientific critics backfired against them, and they ended up being discredited, and he vindicated.

The rest of the history of the defense of Galileo regarding the earth's motion is essentially equivalent to an aspect of the story we have already told above (Chapter 11.1). This is the strand of the historical repercussions involving the discoveries of Newton, Bradley, Guglielmini, Bessel, and Foucault. However, long before Foucault, as it was becoming clearer that Galileo had been right in holding that the earth moves, another genre of clerical apologia had been emerging. Galileo started being charged with believing what turned out to be true for the wrong reasons or with the support of inadequate evidence.

For example, even during Galileo's lifetime, his geokinetic argument from tides had seemed not completely convincing. Then, after Newton's correct explanation of the tides as caused by the gravitational attraction of the moon (and also of the sun), one could also claim that there was definitely an error in Galileo's theory that the tides were caused by the earth's motion. And so the anti-Galilean critics started to mention the tidal argument as one of Galileo's bad reasons for believing what turned out to be true. Today this criticism continues to be one of the most common charges against Galileo.[2]

In 1841, an anonymous article in a German journal inaugurated this kind of apologia in an explicit manner. It argued that the

Inquisition rendered a service to science by condemning the Copernican theory when it had not yet been demonstrated to be true, and by condemning Galileo for supporting it with scientifically incorrect arguments. The critic claimed that the mechanical objections to the earth's motion depended crucially on the assumption that air has no weight; that therefore they could not be answered until the discovery that air has weight; that Galileo was not aware of this fact; and that the discovery was made after his death by Torricelli and Pascal.

This position was historically untenable insofar as Galileo was clearly aware that air has weight.[3] The position was also scientifically misconceived because most of the mechanical difficulties depended not on the weight of air, but on such principles as conservation and composition of motion and inertia, as we saw earlier (Chapter 5). However, this type of criticism raises a crucial and valid point: there is more to being right than that one's beliefs happen to be true, i.e., correspond to reality; it is also important that one's own supporting reasons and evidence are right. In short, one's reasoning is at least as important as the substantive content of one's beliefs. Still, most such anti-Galilean charges can be refuted; Galileo's reasoning can be successfully defended; indeed it can be shown to be a model of critical thinking (see Finocchiaro 1980, 1997, 2010b).

In any case, other issues were bound to arise in the process of coming to terms with the condemnation of Galileo. They involved the principle that Scripture is not a scientific authority. This methodological and theological principle is much more elusive than the astronomical claim that the earth moves, and so the corresponding issues are more complex.

At first, some anti-Galilean critics mentioned this principle as one of Galileo's main errors. For example, in 1651 the already-mentioned Riccioli, besides criticizing the geokinetic theory scientifically, elaborated explicitly a very conservative version of biblical fundamentalism, according to which the literal meaning of biblical statements must be held to be physically true and scientifically correct; thus, allegedly, the Inquisition had been wise in upholding the fundamentalist view against Galileo (cf. Finocchiaro 2005b: 82–84).

Eventually, however, it turned out that Galileo was right regarding this principle as well. As we saw earlier (Chapter 11.1), this occurred with his implicit vindication by Pope Leo XIII's *Providentissimus Deus* in 1893, and with his explicit vindication by Pope John Paul II's rehabilitation of Galileo in 1979–92.

However, once again, as it became increasingly clear that Galileo's hermeneutical principle was correct, his critics started to emphasize the reasons and arguments he had given to justify it. They tried to find all sorts of incoherences in his reasoning. For example: that his essays on the topic contain not only assertions denying the scientific authority of Scripture, but also assertions affirming it (McMullin 1998, 2005b); that he objects to the use of biblical passages against his own astronomical claims, but also tries to interpret the Joshua (10: 12–13) passage in geokinetic terms (McMullin 2005b: 101–2, 110–11; Biagioli 2006: 219–59); that he tries to illegitimately shift the burden of proof by a "sleight of hand ... [to the effect that] it is no longer Galileo's task to prove the Copernican system, but the theologians' task to disprove it" (Koestler 1959: 436–37); and that he wants both to appeal to the theological tradition (e.g., by frequent quotations from St. Augustine) and to overturn it by a radically new principle. Again, it is important to know about the possibility of raising such objections and to understand them, but Galileo can be defended from this criticism of his reasoning, for the criticism is itself criticizable as mistaken.

On the other hand, the greater complexity of the scriptural issue created new possibilities for anti-Galilean criticism. Independently of the truth or falsity of the principle denying the scientific authority of Scripture, and independently of the correctness or incorrectness of Galileo's supporting reasoning, he is sometimes criticized for his theological intrusion and pastoral imprudence. The criticism of theological intrusion objects that Galileo was not a professional theologian, and so he had no right to interfere in hermeneutical discussions. One reply to this criticism is that Galileo did stay away from theological discussions until his scientific ideas were attacked on scriptural grounds; after that, he had every right to defend himself by refuting those attacks as fully as he did. The criticism of pastoral imprudence objects that it was irresponsible for Galileo to loudly proclaim to the popular masses the limitations of the literal interpretations of Scripture at a time when the Catholic Church was in a vital struggle with the Protestant Reformers, given that scriptural interpretation was a key aspect of that struggle.

Something even stranger occurred with regard to the hermeneutical issue. At one point he was blamed for holding and doing the *opposite* of what he actually held and did; i.e., that he preached and practiced the principle that biblical passages should be used to confirm astronomical theories. This criticism started in 1784–85 and was widely accepted for more than a century. It became a slogan: "Galileo was condemned not for being a good astronomer but for being a bad theologian" (cf. Finocchiaro 2005b: 155–57, 159–63; 2010b: 251–76).

Another strand of anti-Galilean criticism focuses on his alleged legal culpability. It claims that the trial did not really deal with the just discussed astronomical-geokinetic or hermeneutical-methodological issues. Galileo was condemned neither for being a good astronomer, nor for being a bad theologian, but rather for something else—disobedience or insubordination. His crime was the violation of the ecclesiastical admonition which he received in February 1616. Admittedly, it is uncertain whether this admonition amounted simply to a warning by Cardinal Bellarmine not to defend the earth's motion, or to the Inquisition's more stringent special injunction not to discuss the topic in any way whatsoever. However, in either case, Galileo's *Dialogue* violated the admonition. The violation of the special injunction not to discuss is clear and direct. And a violation of Bellarmine's warning not to defend can be claimed to have occurred because the book does defend the earth's motion by criticizing the arguments against it and endorsing some in favor.

Such criticism can be dated as far back as 1793, and it continues to be repeated and embellished (for details, see Finocchiaro 2005b: 164–74). Whether correct or incorrect, this criticism is relevant and cannot be summarily dismissed. In my opinion, however, it is untenable. Two distinct points need to be made here.

First, there is the special injunction (not to discuss), from which viewpoint it would seem that Galileo can have no defense.

It turns out, however, that this time the relevant defense is contained in the original documents of the Inquisition proceedings, which have been miraculously preserved and make up a special Galilean file now held at the Vatican Secret Archives. This was discovered in the decade 1867–78, when these proceedings were opened to scholars and published in their entirety. A consensus emerged that the special-injunction document has enough irregularities that this aspect of the proceedings must be regarded as embodying a legal impropriety. From this perspective, the legal criticism of Galileo also backfired against the critics. It emerged that he was the victim of an injustice in a way that had been previously unsuspected. One could almost say that the trial documents suggest that he was framed.

There remains, of course, the criticism that Galileo violated Bellarmine's milder warning (not to defend the geokinetic idea). A possible answer to this criticism is this. As previously discussed (Chapter 8.2), the *Dialogue* discusses the earth's motion by examining all the arguments on both sides; the examination includes not only a presentation and analysis of the arguments, but also their evaluation. Galileo was indeed taking the liberty of *evaluating* the arguments; he was hoping that if he carried out the evaluation correctly, his having engaged in argument evaluation would not be held against him. He was taking the gamble that a correct assessment of arguments would not be seen as an objectionable defense of Copernicanism.

Although such a defense of Galileo has never, to my knowledge, been fully articulated, traces of it can be found in the historical aftermath. In 1943, Pio Paschini, himself a clergyman, explicitly formulated such a defense of Galileo by stating that "it was not his [Galileo's] fault if the arguments for the heliocentric system turned out to be more convincing."[4]

Moreover, there is another issue to be raised with regard to Galileo's alleged disobedience of Bellarmine's warning. Was that warning legitimate? I know of no convincing argument justifying its legitimacy.[5] It may have been one of the many abuses of power in this story. If the warning was not legitimate, then Galileo disobeyed an illegal order. And even if the warning was proper from the point of view of canon law, we may ask whether it was also proper from the *moral* point of view. Again, at worst Galileo may have committed a legal "misdemeanor" while pursuing a morally desirable aim or exercising a basic human right.

One might think that the implicit theological vindication of Galileo by an influential pope in 1893, coming soon after his judicial rehabilitation by the meticulous scholarship of the 1870's, on top of the older and more gradual scientific vindication provided by the proofs of the earth's motion climaxing with Foucault's pendulum (in 1851), that such developments would discourage further indictments of the victim. But to think so would be to underestimate the power of human ingenuity or the unique complexity of the Galileo affair. In fact, a novel apologia was soon devised by a great scholar who combined knowledge of physics, history, and philosophy—Pierre Duhem. In 1908 he advanced the new charge that Galileo was a bad epistemologist.

The criticism of Galileo as a bad epistemologist should not be confused with the criticism that he was a bad arguer, doing a poor job in justifying the truth of the earth's motion and the denial of the scientific authority of Scripture. The epistemological criticism of Galileo attributes to him untenable epistemological or methodological principles and practices, and then it connects such epistemological errors with the trial.

The epistemological doctrine which Duhem found especially objectionable is "realism": that science aims at the truth about the world, and scientific theories are descriptions of physical reality that are true, probably true, or potentially true. Duhem was an advocate of epistemological "instrumentalism": scientific theories are merely instruments for making mathematical calculations and observational predictions, and not descriptions of reality, and so they are not the sort of things that can be true or false, but only more or less convenient. Duhem tried to blame Galileo's trial on epistemological realism, which was allegedly shared by Galileo and his Inquisitors, and also on their joint failure to appreciate instrumentalism, which in that historical context was being allegedly advocated by Cardinal Bellarmine and Pope Urban VIII. In Duhem's own memorable words: "logic was on the side of ... Bellarmine, and Urban VIII, and not on the side of Kepler and Galileo; ... the former had understood the exact import of the experimental method; and ... in this regard, the latter were mistaken" (Duhem 1908: 136; cf. Duhem 1969: 113). To avoid being misled, here Duhem's "logic" should be taken to mean "epistemology," and not reasoning.

Duhem's epistemological criticism of Galileo is interesting and important. Nevertheless, it is untenable, primarily because under the heading of Galilean realism Duhem subsumes too many other epistemological principles besides the ideal of truth and description of reality; but these other attributions are conceptually arbitrary and textually inaccurate. Moreover, Duhem failed to appreciate that Galileo's confirmation of the Copernican theory was simultaneously a confirmation of epistemological realism (and hence a disconfirmation of instrumentalism).

Next, there is the issue of whether Galileo is to be credited or blamed for helping us understand that science and religion are incompatible, or harmonious, as the case may be. But my approach to this problem should be clear from what has already been discussed above, under the heading of the potential lesson regarding the relationship between science and religion.

Finally, it is worth highlighting the issues of recent major developments, because they happen to underscore the need of a Galilean approach to the ongoing Galileo affair, especially the ideals of judiciousness, open-mindedness, and fair-mindedness.

To see this, recall that in 1942 there was the first informal rehabilitation of Galileo by the Church. Let us also recall that, in 1979, Pope John Paul II began a further informal rehabilitation of Galileo that was not concluded until 1992. However, at about the same time, he became the target of unprecedented criticism on the part of various representatives of secular culture (cf. Finocchiaro 2005b: 295–317; 2010b: xxxi–xxxvii, 211–16). It was as if a reversal of roles was occurring, with his erstwhile enemies turning into friends, and his former friends becoming enemies. Several other circumstances add significance to such a development. These critics elaborated what might be called social and cultural criticism of Galileo; that is, they tried to blame Galileo by holding him personally or emblematically responsible for such things as the abuses of the industrial revolution, the social irresponsibility of scientists, the atomic bomb, and the rift between the two cultures (science and the humanities). They were mostly writers with sympathies subsumable under the left wing of the political spectrum. The most outstanding of these critics were central-European German-speaking personalities: Bertolt Brecht was a German playwright who authored the play *Galileo*, which went through three versions (1938, 1947, 1955) and became a classic of twentieth-century theater; Arthur Koestler was a Hungarian-born novelist and intellectual who, in 1958, published *The Sleepwalkers: A History of Man's Changing Vision of the Universe*, which became an international best-seller; and Paul Feyerabend was an Austrian-born philosophy professor at the University of California, Berkeley, who advanced his version of social criticism in *Against Method*, which also went through three editions (1975, 1988, 1993).

These developments have not been properly assimilated yet. For example, the Catholic "rehabilitations" tend to be either unfairly criticized (even by Catholics), or uncritically accepted (even by non-Catholics). Moreover, Pope Benedict XVI seems to have displayed an ambivalent attitude toward this issue; his ambivalence is revealing, but continues to polarize. And the left-leaning social critiques tend to be summarily dismissed by practicing scientists, whose professional identity is thereby threatened, or dogmatically advocated by self-styled progressives, who apparently have not learned much from Galileo and want to turn the clock back to pre-Galilean days. In any case, these developments clearly suggest that the Galileo affair transcends religion, and has become not so much a controversy over Church affairs, but a problem in human behavior and society generally.

To summarize, the Copernican revolution required that the geokinetic hypothesis be justified not only with new theoretical arguments but also with new observational evidence; that the earth's motion be not only supported constructively, but also critically defended from many powerful old and new objections; and that this defense include not only the destructive refutation but also the appreciative understanding of those objections in all their strength. One of Galileo's major accomplishments was not only to provide new evidence supporting the earth's motion, but also to show how those objections could be refuted, and to elaborate their power before they were answered. In this sense, Galileo's defense of Copernicus was rational-minded, open-minded, fair-minded, and judicious.

Moreover, we have also seen that an essential thread of the subsequent Galileo affair has been the emergence of many anti-Galilean criticisms, from the point of view of astronomy, physics, theology, hermeneutics, logic, epistemology, methodology, law, morals, and social awareness. In my account just sketched, I have made it clear both that such criticisms arise naturally and legitimately, and that Galileo has been, and can be, effectively defended from them. Accordingly, I claim that the proper defense of Galileo should have the reasoned, judicious, open-minded, and fair-minded character which his own defense of Copernicanism had. This thesis holds the key for the resolution of the ongoing Galileo affair.

This is a normative thesis that amounts to saying that defending Copernicanism in the reasoned and judicious manner in which Galileo did is instructive and suggests the proper way in which Galileo himself can and should be defended from the many attempts to justify his condemnation. This is a lesson that results if, besides trying to understand what really happened in the Copernican revolution and the Galileo affair, we also try to assess what is right from various nuanced points of view; in particular, if we try to learn from Galileo. And the lesson is that just as Galileo's defense of Copernicus owed its success to its being reasoned and judicious, so our defense of Galileo can succeed if it possesses the same qualities.

To defend Galileo in this manner does not mean to show that he was completely and always right; it only means to show that he was essentially right, or more nearly correct than not. Such a defense of Galileo is not an attempt to show that criticisms of him are without foundation; rather, the defense cannot even get started unless one first knows and understands that there are reasons for attributing to him various errors or improprieties; in such a context one tries to show that such anti-Galilean arguments are ultimately invalid, or at least weaker than the pro-Galilean ones. Defending Galileo is not meant to be a one-sided exercise pointing out only his merits and virtues; rather merits and virtues are meant to be inherently comparative properties whose positive aspects are seen only vis-à-vis the negative ones. Nor is the defense of Galileo a hagiographic exercise exaggerating the number or importance of his scientific and cultural achievements; in this regard, I want to stress that my position has a historical and interpretive component, besides the philosophical and evaluative one, and that a main thrust of the interpretive component is the historical reality of the anti-Galilean criticisms.

Thus, in my view, on the one hand the proposition that Galileo's defense of Copernicanism was wrong (i.e., that Galileo's condemnation was right) is almost as false and untenable as the proposition that the earth stands still at the center of the universe. On the other hand, the arguments purporting to justify various Galilean improprieties are in appearance almost as plausible as the anti-Copernican arguments seemed to be in the sixteenth century. But, ultimately, the anti-Galilean arguments can be shown to be almost as weak as the anti-Copernican arguments were shown to be by Galileo.

NOTES

1 Galluzzi 1977, Koyré 1955; cf. Finocchiaro 2001: 500.
2 See, e.g., Graney 2010: 14; Heilbron 2010: 116, 216–17, 260; Shea 1972: 172–89, 2005.
3 Galilei (1890–1909, vol. 8: 123–24, vol. 12: 33–36, vol. 14: 158; 1974: 82–83).
4 Paschini 1943: 97; cf. Finocchiaro 2005b: 280–84, DiCanzio 1996: 309.
5 But see the important technicalities discussed in Mereu 1979: 435–37, Beretta 1998: 239–48.

APPENDIX
TABLE OF CROSS-REFERENCES AMONG EDITIONS

The following table contains cross-references between the four important editions of Galileo's *Dialogue* cited in this book: Galilei 1897 (abbreviated FAV, for the editor Favaro); Galilei 1967 (abbreviated DCA, for the translator Drake and the publisher, the University of California Press); Galilei 1997 (abbreviated FIN, for the translator Finocchiaro); and Galilei 2001 (abbreviated DML, for the translator Drake and the publisher, the Modern Library). Its main purpose is to avoid constantly having to give four sets of page references in the course of my exposition and notes; this would have taxed readers' eyes and attention to an undesirable degree. On the other hand, the reasons for providing four sets of references, rather than just one or two, are as follows.

In these two editions of Drake's translation, the English text is identical, but they differ in pagination. However, the 2001 Modern Library edition seems a more user-friendly book and costs much less than the 1967 California edition. For these reasons, I have chosen Galilei 2001 as the main edition for giving references throughout my exposition and notes. Thus, readers will always find references to this main edition (DML = Galilei 2001) whenever I mention Galilean passages, whether or not I am quoting from other editions. However, for at least two generations, the California edition of Drake's translation has been the standard English translation, used by countless readers in their education and scholarly work, including the present writer. For this reason, I thought it valuable to give cross-references to Galilei 1967 (= DCA). Moreover, there are obvious reasons for wanting to have cross-references to Galileo's Italian text, as found in the seventh volume of the critical edition of his collected works (Galilei 1897 = FAV). Similarly, in 1997, I published my own abridged translation of the *Dialogue*, and although it does not include the entire text, I believe it is more accurate from the points of view of the meaning of Galileo's words and the structure of his language; thus, it ought to come as no surprise that, whenever possible, I quote from Galilei 1997 (= FIN).

The table has four columns, corresponding to these four editions. Each column lists the page numbers and ranges on which we can find all the particular Galilean passages mentioned, discussed, or quoted in this book. Each row (or line) in the table starts with the page number(s) where a referenced passage appears in DML, and then gives the corresponding page number(s) in DCA, FAV, and FIN. Because, as mentioned, Galilei 1997 is an abridged translation, only about half the slots in its column are filled. In some cases, distinct passages from the same DML page (e.g., 37, 271) appear on more than one page in the other editions, and so some lines in the table display partial overlap; however, there is no real duplication and the references given in the course of my exposition make clear that different passages are involved. Finally, note that the decimal numbers in the FAV references denote line numbers.

Galilei 2001 (= DML)	Galilei 1967 (= DCA)	Galilei 1897 (= FAV)	Galilei 1997 (= FIN)
5–7	5–7	29–31	77–82
6	6	30	79–80
9–15	9–14	33–38	
11	11	35	

(continued)

Galilei 2001 (= DML)	Galilei 1967 (= DCA)	Galilei 1897 (= FAV)	Galilei 1997 (= FIN)
15	14	38	
15–20	14–18	38–42	
20	19	42–43	
20–36	18–32	42–57	
21	19–20	43.12–30	
21–22	20	43.40–44.8	
22–23	20	44.8–45.10	
23–32	21–28	45.11–53.6	
25	23	47	
32–35	29–31	53.13–55.1	
35	31	55.3–4, 56.1–9	
35–36	31–32	56.9–20	
36	31–32	56.20–25	
36–37	32–33	57	83–84
36–43	32–38	57–62	83–90
37	33	57.20–27	
37	33	57.27–34, 58.5–8	
37	33	57.34–58.4	
37–38	33	58.5–8	
38–39	34	58–59	86
40	35–36	60.10–24	
40–41	36	60.25–61.15	
43–44	38–39	63.18–64.11	
43–53	38–47	62–71	
44	39	63–64	
44–46	39–41	64.11–65.24	
46–48	41–42	65.26–67.18	
48–51	42–44	67.19–69.24	
51–52	45–46	69.25–71.9	
53–54	47	71–72	91–92
53–66	47–58	71–83	91–107
54	47–48	72	91–92
54–56	48–49	72.31–74.11	
56	49	74.12–23	
56–57	49–50	74.23–75.8	
57	50	75	96
57–58	50–51	75.9–76.11	
58	50–51	75–76	96–98
61	53–54	78	101
63	55–56	80	104

(continued)

Galilei 2001 (= DML)	Galilei 1967 (= DCA)	Galilei 1897 (= FAV)	Galilei 1997 (= FIN)
63–65	55–57	80.16–82.7	
67–71	58–62	83–87	
68	59	84	
68–69	60	84–85	
69–70	60–61	85	
71	62	87.18–21	
71–72	62–63	87	
71–82	62–71	87–96	
82–84	71–73	96–98	
82–100	71–87	96–112	
86–87	75–76	100–101	
92	80	105	
95–96	83	108–9	
96	84	109	
100–103	87–90	113–15	
100–113	87–98	112–24	
102–3	89–90	115	
103–4	90–91	115–16	
104–5	91	116–17	
105–110	91–96	117–21	
112–13	97–98	123–24	
113	97–98	124	
113–21	98–105	124–31	107–16
118	103	129	
118–21	103–5	128–31	113–16
123–32	106–14	132–39	117–28
123–55	106–33	132–59	117–55
131	113	138–39	127
132–44	114–24	139–50	128–42
144	124	150	142–43
144–45	124–25	150	143
144–55	124–33	150–59	142–55
146–47	126–27	152–53	146
147	127	153	146
147–48	126–27	152–53	146–47
148	127–28	153–54	147–48
148–53	127–31	153–58	147–53
151	130	156	151–52
153–54	132	158–59	154
155	133	159	

(continued)

Galilei 2001 (= DML)	Galilei 1967 (= DCA)	Galilei 1897 (= FAV)	Galilei 1997 (= FIN)
155	134	160	
155–58	133–36	159–62	
158	136	162	
158–61	136–38	162–64	
159–61	137–38	163–64	
160	138	164	
161	139	164–65	155–56
161–64	138–41	164–67	155–58
164–66	141–43	167–69	158–62
164–79	141–54	167–80	158–70
166–79	143–54	169–80	162–70
167	144	169–70	163
173–79	149–54	175–80	
179	154	180	
179–95	154–68	180–94	
191–92	165	191	
192	166	192	
194	167	193	
195–96	168	194	
195–98	168–71	194–97	
198	170–71	196	
198	171	196–97	
198–206	171–78	197–203	
199	171	197	
202	174	200	
206–8	178–80	203–5	
208	179	205	
208–12	180–83	205–9	
209	180	205–6	
212–13	183	209	
212–18	183–88	209–14	
213–14	183–84	209.26–210.15	
214	184	210.16–20	
216	186	212.3–10	
216	186	212.10–30	
216–18	186–88	212–14	
218	188	214	171
218–20	188–90	214–16	171–73
218–53	188–218	214–44	171–212
219–20	189	215–16	172

(continued)

Galilei 2001 (= DML)	Galilei 1967 (= DCA)	Galilei 1897 (= FAV)	Galilei 1997 (= FIN)
220	190	216	173
222–29	190–97	217–23	175–82
226–27	195	221	179–80
229–36	197–203	223–29	182–93
230	198	224	184
231	199	225	186
236	203	229	193
236–41	203–7	229–33	193–99
236–44	203–10	229–37	193–202
245–53	211–18	237–44	203–12
250	215	242	208–9
250	216	242	209
253–71	218–33	244–60	
254–55	218–20	245–46	
259	223	250	
271	233	259–60	
271	233–34	260	
271–72	233–34	260	
271–86	233–47	260–72	
272	234–35	260–61	
277	239	265	
282	243	268.16–24	
287–98	247–57	272–81	212–20
288–90	248–49	273–74	213–14
288–95	248–55	273–79	212–17
294–95	253–54	278–79	215–16
296–97	254–55	279–80	217–18
296–98	255–56	280–81	218–19
298–99	257–58	281–82	
298–306	257–64	281–88	
304–5	262	287	
305–6	263	288	
307–8	264–66	289–90	
307–19	264–75	289–98	
309	266	290–91	
309–10	266–67	291	
310–12	267–69	292–93	
311–12	268–69	292–93	
312–14	269–71	293–95	
312–17	269–73	293–97	

(continued)

Galilei 2001 (= DML)	Galilei 1967 (= DCA)	Galilei 1897 (= FAV)	Galilei 1997 (= FIN)
321–70	276–318	299–346	
323	277	300	
370–80	318–27	346–55	221–33
379–80	326–27	354	232–33
380	327	355	233
381	327–28	355–56	234–35
381–95	327–40	355–68	234–44
388	334	356–62	235–37
390	335	363	238
396–400	340–45	368–72	
397	342	370	
399	343	371	
400	344	372	
400–401	345	372	
400–414	345–56	372–83	
404	348	376	
410	353	380	
412	355	382	
413	356	383	
413–14	356	383	
414–32	356–72	383–99	245–64
415	357	384	246
416	358	385	248
426	367	394	258–59
430	370	397	262
432–61	372–97	399–423	264–81
433–36	373–76	400–3	
436–37	376–77	403–4	
439–40	378–79	405–6	
445–46	383–85	410–11	271–74
447	385	412	274–75
448	386	413	276
448–50	386–88	413–14	277–79
449	387	413–14	278
452–61	389–97	416–23	
461–62	398	424	
461–81	397–415	423–41	
462	398	424	
462–63	398–99	424–25	
479	413	439	

(continued)

Galilei 2001 (= DML)	Galilei 1967 (= DCA)	Galilei 1897 (= FAV)	Galilei 1997 (= FIN)
483–92	416–24	442–50	
487	420	446	283
488	420	446	284
490–92	422–24	448–50	286–88
492	423–24	450	288
492	424	450	288
492–506	424–36	450–62	288–303
502	432–33	458–59	298–99
506–16	436–44	462–70	
509	438	464	
511	440	465–66	
514	443	468	
516–35	444–61	470–85	
533–35	460–61	484–85	
535–36	462	486	304
535–39	461–65	485–89	303–8
537–39	463–65	487–89	305–8

SELECTED BIBLIOGRAPHY

Aiton, E.J. (1954) "Galileo's Theory of the Tides," *Annals of Science*, 10: 44–57.

——(1963) "On Galileo and the Earth-Moon System," *Isis*, 54: 265–66.

——(1965) "Galileo and the Theory of the Tides," *Isis*, 56: 56–61.

Aristotle (1952) *Works of Aristotle*, 2 vols, in *Great Books of the Western World*, vols 8–9, Chicago: Encyclopedia Britannica.

Barker, P., and B.R. Goldstein (1998) "Realism and Instrumentalism in Sixteenth Century Astronomy," *Perspectives on Science*, 6: 232–58.

Beltrán Marí, A. (2006) *Talento y poder*, Pamplona: Laetoli.

Beretta, F. (1998) *Galilée devant le Tribunal de l'Inquisition*, Unpublished dissertation, Faculty of Theology, University of Fribourg, Switzerland.

Biagioli, M. (1993) *Galileo Courtier*, Chicago: University of Chicago Press.

——(2006) *Galileo's Instruments of Credit*, Chicago: University of Chicago Press.

Blackwell, R.J. (1991) *Galileo, Bellarmine, and the Bible*, Notre Dame: University of Notre Dame Press.

Brahe, Tycho (1596) *Epistolae astronomicae*, Uraniborg.

——(1602) *Astronomiae instauratae progymnasmata*, Uraniborg.

Brown, H.I. (1976) "Galileo, the Elements, and the Tides," *Studies in History and Philosophy of Science*, 7: 337–51.

——(1977) *Perception, Theory and Commitment*, Chicago: Precedent.

Bucciantini, M. (2003) *Galileo e Keplero*, Turin: Einaudi.

Bucciantini, M. and M. Camerota (eds) (2009) *Scienza e religione*, Rome: Donzelli.

Burstyn, H.L. (1962) "Galileo's Attempt to Prove that the Earth Moves," *Isis*, 53: 161–85.

——(1963) "Galileo and the Earth-Moon System," *Isis*, 54: 400–401.

——(1965) "The Deflecting Force of the Earth's Rotation from Galileo to Newton," *Annals of Science*, 21: 47–80.

Camerota, M. (2004) *Galileo Galilei e la cultura scientifica nell'età della Controriforma*, Rome: Salerno Editrice.

Campanella, T. (1994) *A Defense of Galileo, the Mathematician from Florence*, trans. and ed. R.J. Blackwell, Notre Dame: University of Notre Dame Press.

Chalmers, A., and R. Nicholas (1983) "Galileo and the Dissipative Effects of a Rotating Earth," *Studies in History and Philosophy of Science*, 14: 315–40.

Chiaramonti, Scipione (1628) *De tribus novis stellis quae annis 1572, 1600, 1604 comparuere*, Cesena.

Clutton-Brock, M., and D. Topper (2011) "The Plausibility of Galileo's Tidal Theory," *Centaurus*, 53: 221–35.

Cohen, H.F. (1994) *The Scientific Revolution*, Chicago: University of Chicago Press.

Cohen, I.B. (1960) *The Birth of a New Physics*, Garden City: Doubleday.

——(1967) "Newton's Attribution of the First Two Laws of Motion to Galileo," in *Atti del symposium internazionale di storia, metodologia, logica e filosofia della scienza*: xxv-xliv, Vinci: Gruppo italiano di storia della scienza.

——(1999) "A Guide to Newton's *Principia*," in Newton 1999: 1–370.

Conti, L. (1990) "Francesco Stelluti, il copernicanesimo e la teoria galileiana delle maree," in C. Vinti (ed.) *Galileo e Copernico*: 141–236, Perugia: Porziuncola.

Copernicus, Nicolaus (1952) *On the Revolutions of the Heavenly Spheres*, in *Great Books of the Western World*, vol. 16: 481–844, Chicago: Encyclopedia Britannica.

——(1992) *On the Revolutions*, trans. and ed. E. Rosen, Baltimore: Johns Hopkins University Press.

Danielson, D.R. (2009) "Myth 6: That Copernicanism Demoted Humans from the Center of the Cosmos," in R.L. Numbers (ed.) *Galileo Goes to Jail and Other Myths about Science and Religion*: 50–58, Cambridge: Harvard University Press.

Descartes, René (1897–1913) *Oeuvres*, 13 vols., eds C. Adam and P. Tannery, Paris: Cerf.

DiCanzio, A. (1996) *Galileo: His Science and His Significance for the Future of Man*, Portsmouth: Adasi.

Drake, S. (1970) *Galileo Studies*, Ann Arbor: University of Michigan Press.

——(1978) *Galileo at Work*, Chicago: University of Chicago Press.

——(1979) "History of Science and Tide Theories," *Physis*, 21: 61–69.

——(1983) *Telescopes, Tides & Tactics*, Chicago: University of Chicago Press.

——(1986) "Reexamining Galileo's *Dialogue*," in Wallace 1986: 155–75.

——(1990) *Galileo: Pioneer Scientist*, Toronto: University of Toronto Press.

Drake, S., and C.D. O'Malley (trans and eds) (1960) *The Controversy on the Comets of 1618*, Philadelphia: University of Pennsylvania Press.

Draper, J.W. (1875) *History of the Conflict between Religion and Science*, New York: Appleton.

Dreyer, J.J.E. (1953) *A History of Astronomy from Thales to Kepler*, 2nd edn, New York: Dover.

Duhem, P. (1908) *SOZEIN TA PHAINOMENA: Essai sur la notion de theorie physique de Platon à Galilée*, Paris: Hermann.

——(1969) *To Save the Phenomena*, trans E. Doland and C. Maschler, Chicago: University of Chicago Press.

Einstein, A. (1953) "Foreword," in Galilei 1953a: vi–xx.

——(1954) *Ideas and Opinions*, trans. S. Bargmann, New York: Crown.

Ennis, R.H. (1996) "Critical Thinking Dispositions," *Informal Logic*, 18: 165–82.

Fantoli, A. (2003) *Galileo: For Copernicanism and for the Church*, 3rd revised edn, trans. G.V. Coyne, Vatican City: Vatican Observatory.

——(2012) *The Case of Galileo*, trans. G.V. Coyne, Notre Dame: University of Notre Dame.

Feyerabend, P.K. (1988) *Against Method*, revised edn, London: Verso.

Finocchiaro, M.A. (1980) *Galileo and the Art of Reasoning: Rhetorical Foundations of Logic and Scientific Method*, Boston Studies in the Philosophy of Science, vol. 61, Dordrecht: Reidel [now Springer].

——(1985) "Wisan on Galileo and the Art of Reasoning," *Annals of Science*, 42: 613–16.

——(trans. and ed.) (1989) *The Galileo Affair: A Documentary History*, Berkeley: University of California Press.

——(trans. and ed.) (1997) *Galileo on the World Systems: An Abridged Translation and Guide*, Berkeley: University of California Press.

——(2001) "Aspects of the Controversy about Galileo's Trial (from Descartes to John Paul II)," in J. Montesinos and C. Solís (eds) *Largo Campo di Filosofare*: 491–512, La Orotava: Fundación Canaria Orotava de Historia de la Ciencia.

——(2003) "Physical-Mathematical Reasoning," *Synthese*, 134: 217–44.

——(2005a) *Arguments about Arguments*, Cambridge: Cambridge University Press.

——(2005b) *Retrying Galileo, 1633–1992*, Berkeley: University of California Press.

——(2010a) "Defending Copernicus and Galileo: Critical Reasoning and the Ship Experiment Argument," *Review of Metaphysics*, 64: 75–103.

——(2010b) *Defending Copernicus and Galileo: Critical Reasoning in the Two Affairs*, Boston Studies in the Philosophy of Science, vol. 280, Dordrecht: Springer.

——(2011) "Fair-mindedness vs. Sophistry in the Galileo Affair," in M. Dascal and V. Boantza (eds) *Controversies Within the Scientific Revolution*: 53–73, Amsterdam: John Benjamins.

Fisher, A. (1991) "Testing Fairmindedness," *Informal Logic*, 13: 31–36.

Fisher, A., and M. Scriven (1997) *Critical Thinking*, Point Reyes: Edgepress.

Freud, S. (1922) *Introductory Lectures on Psycho-Analysis*, trans. J. Riviere, London: George Allen and Unwin.

Galilei, Galileo (1632) *Dialogo sopra i due Massimi Sistemi del mondo, tolemaico e copernicano*, Florence.

——(1635) *The Dialogues of Galileus Galilei*, trans. Joseph Webbe, unpublished English translation of Galileo's *Dialogue*, London: British Library, manuscript number Harleian MS 6320.

——(1661) *System of the World*, trans. Thomas Salusbury, in T. Salusbury (trans. and ed.) *Mathematical Collections and Translations*, 2 vols, London: Leybourne, 1661–65, vol. 1, part 1: 1–424.

——(1890–1909) *Le Opere di Galileo Galilei*, 20 vols, National Edition by Antonio Favaro et al., Florence: Barbèra.

——(1891) *Dialog über die beiden hauptsächlichsten Weltsysteme, das ptolemäische und das kopernikanische*, trans. and ed. Emil Strauss, Stuttgart: Teubner.

——(1897) *Dialogo sopra i due massimi sistemi del mondo, tolemaico e copernicano*, National Edition by Antonio Favaro et al., Florence: Barbèra (= Galilei 1890–1909, vol. 7).

——(1953a) *Dialogue Concerning the Two Chief World Systems, Ptolemaic and Copernican*, trans. and ed. S. Drake, Berkeley: University of California Press.

——(1953b) *Dialogue on the Great World Systems*, Salusbury's translation revised by Giorgio de Santillana, Chicago: University of Chicago Press.

——(1964) *Opere*, 5 vols, ed. Pietro Pagnini, Florence: Salani.

——(1967) *Dialogue Concerning the Two Chief World Systems, Ptolemaic and Copernican*, trans. and ed. Stillman Drake, 2nd revised edn, Berkeley: University of California Press.

——(1970) *Dialogo sopra i due massimi sistemi del mondo, tolemaico e copernicano*, ed. Libero Sosio, Turin: Einaudi.

——(1974) *Two New Sciences*, trans. and ed. S. Drake, Madison: University of Wisconsin Press.

——(1982), *Dialog über die beiden hauptsächlichsten Weltsysteme, das ptolemäische und das kopernikanische*, trans. E. Strauss (1891), eds R. Sexl and K. von Meyenn, Stuttgart: Teubner.

——(1992) *Dialogue sur les Deux Grands Systèmes du Monde*, trans and eds René Fréreux and François de Gandt, Paris: Seuil.

——(1994) *Diálogo sobre los dos máximos sistemas del mundo, ptolemaico y copernicano*, trans. and ed. Antonio Beltrán Marí, Madrid: Alianza Editorial.

——(1997) *Galileo on the World Systems: A New Abridged Translation and Guide*, trans. and ed. M.A. Finocchiaro, Berkeley: University of California Press.

——(1998) *Dialogo sopra i due massimi sistemi del mondo, tolemaico e copernicano*, 2 vols, critical edition and commentary by Ottavio Besomi and Mario Helbing, Padua: Antenore.

——(2001) *Dialogue Concerning the Two Chief World Systems, Ptolemaic and Copernican*, trans. and ed. S. Drake, 2nd revised edn, intro. J.L. Heilbron, New York: Modern Library.

——(2003) *Dialogo sopra i due massimi sistemi del mondo, tolemaico e copernicano*, ed. with commentary by Antonio Beltrán Marí, Milan: Rizzoli.

——(2008) *The Essential Galileo*, trans. and ed. M.A. Finocchiaro, Indianapolis: Hackett.

——(2012) *Selected Writings*, trans and eds W.R. Shea and M. Davie, Oxford: Oxford University Press.

Galilei, Galileo, and Christoph Scheiner (2010) *On Sunspots*, trans and eds Eileen Reeves and Albert Van Helden, Chicago: University of Chicago Press.

Galluzzi, P. (1977) "Galileo contro Copernico," *Annali dell'Istituto e Museo di Storia della Scienza*, 2: 87–148.

——(2000) "Gassendi and l'*Affaire Galilée* of the Laws of Motion," *Science in Context*, 13: 509–45.

Gapaillard, J. (1990–91) "Galilée et l'expérience de Locher," *Sciences et techniques en perspective*, 2: 1–10.

——(1992) "Galilée et le principe du chasseur," *Revue d'histoire des sciences*, 45: 281–306.

——(1993) *Et pourtant elle tourne! Le mouvement de la Terre*, Paris: Seuil.

Gatti, H. (1999) *Giordano Bruno and Renaissance Science*, Ithaca: Cornell University Press.

——(ed.) (2002) *Giordano Bruno: Philosopher of the Renaissance*, Aldershot: Ashgate.

Gaukroger, S. (1978) *Explanatory Structures*, Atlantic Highlands: Humanities.

Geymonat, L. (1965) *Galileo Galilei*. trans. S. Drake, New York: McGraw-Hill.

Gingerich, O. (1982) "The Galileo Affair," *Scientific American*, August, pp. 132–43.

Govier, T. (1999) *The Philosophy of Argument*, Newport News: Vale Press.

Graney, C.M. (2010) "Seeds of a Tychonic Revolution," *Physics in Perspective*, 12: 4–14.

——(2011a) "Contra Galileo," *Physics in Perspective*, 13: 387–400.

——(2011b) "Coriolis Effect, Two Centuries before Coriolis," *Physics Today*, August, pp. 8–9.

Hall, A.R. (1954) *The Scientific Revolution, 1500–1800*, London: Longmans, Green, & Co.

Harris, W.H., and J.S. Levey (eds) (1975) *The New Columbia Encyclopedia*, New York: Columbia University Press.

Hawking, S.W. (1988) *A Brief History of Time*, New York: Bantam Books.

——(2002) "Galileo Galilei (1564–1642): His Life and Work," in Galileo Galilei, *Dialogues Concerning Two New Sciences*, ed. with commentary by S. Hawking: xi–xvii. Philadelphia: Running Press.

Heilbron, J.L. (2001) "Introduction," in Galilei 2001: xii–xxi.

——(2010) *Galileo*, Oxford: Oxford University Press.

Henry, J. (1997) *The Scientific Revolution and the Origins of Modern Science*, New York: St. Martin's Press.

Herivel, J. (1965) *The Background to Newton's Principia*, Oxford: Clarendon Press.

Hill, D.K. (1984) "The Projection Argument in Galileo and Copernicus," *Annals of Science*, 41: 109–33.

Hutchison, K. (1990) "Sunspots, Galileo, and the Orbit of the Earth," *Isis*, 81: 68–74.

Jardine, N. (1991) "Demonstration, Dialectic, and Rhetoric in Galileo's *Dialogue*," in D.R. Kelley and R.H. Popkin (eds), *The Shapes of Knowledge from the Renaissance to the Enlightenment*: 101–22, Dordrecht: Kluwer.

John Paul II (1979) "Deep Harmony which Unites the Truths of Science with the Truths of Faith," *L'Osservatore Romano*, Weekly Edition in English, 26 November, pp. 9–10.

——(1992) "Faith can never conflict with reason," *L'Osservatore Romano*, Weekly Edition in English, 4 November, pp. 1–2.

Johnson, R.H. (2000) *Manifest Rationality*, Mahwah: Lawrence Erlbaum.

Johnstone, H.W., Jr. (1959) *Philosophy and Argument*, University Park: Pennsylvania State University Press.

——(1978) *Validity and Rhetoric in Philosophical Argument*, University Park: The Dialogue Press of Man & World.

Koestler, A. (1959) *The Sleepwalkers*, New York: Macmillan.

Koyré, A. (1955) *A Documentary History of the Problem of Fall from Kepler to Newton*, in *Transactions of the American Philosophical Society*, new series, vol. 45, part 4: 329–95, Philadelphia: American Philosophical Society.

——(1966) *Etudes galiléennes*, Paris: Hermann.

——(1968) *Metaphysics and Measurement*, Cambridge: Harvard University Press.

——(1978) *Galileo Studies*, trans. J. Mepham, Hassocks: Harvester.

Kuhn, T.S. (1957) *The Copernican Revolution*, Cambridge: Harvard University Press.

——(1962) *The Structure of Scientific Revolutions*, Chicago: University of Chicago Press.

——(1970) *The Structure of Scientific Revolutions*, 2nd edn, enlarged, Chicago: University of Chicago Press.

Lakatos, I., and E. Zahar (1975) "Why Did Copernicus' Research Program Supersede Ptolemy's?" in R.S. Westman (ed.) *The Copernican Achievement*: 354–83, Berkeley: University of California Press.

Langford, J.J. (1966) *Galileo, Science and the Church*, Ann Arbor: University of Michigan Press.

Lindberg, D.C. (1992) *The Beginnings of Western Science*, Chicago: University of Chicago Press.

Lindberg, D.C., and R.S. Westman (eds) (1990) *Reappraisals of the Scientific Revolution*, Cambridge: Cambridge University Press.

Locher, Ioannes Georgius (1614) *Disquisitiones mathematicae de controversiis et novitatibus astronomicis*, Ingolstadt.

Mach, E. (1960) *The Science of Mechanics*, trans. T.J. McCormack, La Salle: Open Court.

Machamer, P. (1973) "Feyerabend and Galileo," *Studies in History and Philosophy of Science*, 4: 1–46.

——(ed.) (1998) *The Cambridge Companion to Galileo*, Cambridge: Cambridge University Press.

MacLachlan, J. (1977) "Mersenne's Solution for Galileo's Problem of the Rotating Earth," *Historia Mathematica*, 4: 173–82.

——(1990) "Drake Against the Philosophers," in T.H. Levere and W.R. Shea (eds) *Nature, Experiment, and the Sciences*: 123–44, Dordrecht: Kluwer.

Margolis, H. (1987) *Patterns, Thinking, and Cognition*, Chicago: University of Chicago Press.

——(1991) "Tycho's System & Galileo's *Dialogue*," *Studies in History and Philosophy of Science*, 22: 259–75.

McMullin, E. (ed.) (1967) *Galileo: Man of Science*, New York: Basic Books.

——(1998) "Galileo on Science and Scripture," in Machamer 1998: 271–347.

——(ed.) (2005a) *The Church and Galileo*, Notre Dame: University of Notre Dame Press.

——(2005b) "Galileo's Theological Venture," in McMullin 2005a: 88–116.

Mereu, I. (1979) *Storia dell'intolleranza in Europa*, Milan: Mondadori.

Millman, A.B. (1976) "The Plausibility of Research Programs," in F. Suppe and P.D. Asquith (eds) *PSA 1976: Proceedings of the 1976 Biennial Meeting of the Philosophy of Science Association*, vol. 1: 140–48, East Lansing: Philosophy of Science Association.

Moss, J.D. (1993) *Novelties in the Heavens*, Chicago: University of Chicago Press.

Moss, J.D., and W.A. Wallace (2003) *Rhetoric & Dialectic in the Time of Galileo*, Washington: Catholic University of America Press.

Mueller, P.D. (2000) "An Unblemished Success," *Journal for the History of Astronomy*, 31: 279–99.

Naylor, R. (2007) "Galileo's Tidal Theory," *Isis*, 98: 1–22.

Newton, Isaac (1999) *The Principia: Mathematical Principles of Natural Philosophy*, trans and eds I.B. Cohen and A. Whitman, Berkeley: University of California Press.

Palmieri, P. (1998) "Re-Examining Galileo's Theory of Tides," *Archive for the History of Exact Sciences*, 53: 223–375.

——(2001) "Galileo and the Discovery of the Phases of Venus," *Journal for the History of Astronomy*, 32: 109–29.

——(2008) "Galileus Deceptus, Non Minime Decepit," *Journal for the History of Astronomy*, 39: 425–52.

Paschini, P. (1943) "L'insegnamento di Galileo," *Studium*, April, 39: 94–97.

Perelman, Ch., and L. Olbrechts-Tyteca (1969) *The New Rhetoric*, trans J. Wilkinson and P. Weaver, Notre Dame: University of Notre Dame.

Popper, K.R. (1963) *Conjectures and Refutations*, New York: Harper.

Ptolemy, Claudius (1952) *The Almagest*, in *Great Books of the Western World*, vol. 16: 1–478, Chicago: Encyclopedia Britannica.

Ronan, C.A. (1974) *Galileo*, New York: Putnam's Sons.

Rosen, E. (ed.) (1959) *Three Copernican Treatises*, 2nd edn, New York: Dover.

——(1992) "Commentary," in Copernicus 1992: 331–439.

Russell, B. (1997) *Religion and Science*, New York: Oxford University Press.

Schuster, A. (1916) "The Common Aims of Science and Humanity," in *Report of the Eighty-fifth Meeting of the British Association for the Advancement of Science (Manchester, 1915)*: 3–23, London: John Murray. Available online at <http://www.biodiversitylibrary.org/item/95822> (accessed 10 August 2012).

Scriven, M. (1976) *Reasoning*, New York: McGraw-Hill.

Seeger, R.J. (1966) *Galileo Galilei, His Life and His Works*, Oxford: Pergamon.

Shapere, D. (1984) *Reason and the Search for Knowledge*, Dordrecht: Kluwer.

Shea, W.R. (1972) *Galileo's Intellectual Revolution*, New York: Science History.

——(2005) Review of Finocchiaro's *Retrying Galileo*, *Isis*, 96: 644.

Smith, A.M. (1985) "Galileo' Proof for the Earth's Motion from the Movement of Sunspots," *Isis*, 76: 543–51.

Sobel, D. (1999) *Galileo's Daughter*, New York: Walker & Company.

Speller, J. (2008) *Galileo's Inquisition Trial Revisited*, Frankfurt: Peter Lang.

Swerdlow, N.M. (1998) "Galileo's Discoveries with the Telescope and Their Evidence for the Copernican Theory," in Machamer 1998: 244–70.

Topper, D. (1999) "Galileo, Sunspots, and the Motions of the Earth," *Isis*, 90: 757–67.

——(2000) " 'I Know That What I Am Saying Is Rather Obscure'," *Centaurus*, 42: 288–96.

——(2003) "Colluding with Galileo," *Journal for the History of Astronomy*, 34: 75–76.

Toulmin, S., and J. Goodfield, (1961) *The Fabric of the Heavens*, New York: Harper.

Tredennick, H. (ed.) (1969) *The Last Days of Socrates*, New York: Penguin.

Van Helden, A. (1985) *Measuring the Universe*, Chicago: University of Chicago Press.

Vickers, B. (1983) "Epideictic Rhetoric in Galileo's Dialogo," *Annali dell'Istituto e Museo di Storia della Scienza di Firenze*, 8: 69–102.

Wallace, W.A. (1984) *Galileo and His Sources*, Princeton: Princeton University Press.

——(ed.) (1986) *Reinterpreting Galileo*, Washington: Catholic University of America Press.

——(1992a) *Galileo's Logical Treatises*, Dordrecht: Kluwer.

——(1992b) *Galileo's Logic of Discovery and Proof*, Dordrecht: Kluwer.

Wallis, C.G. (1952) "Notes and Commentary," in Copernicus 1952: 481–844.

Westman, R.S. (2011) *The Copernican Question*, Berkeley: University of California Press.

White, A.D. (1896) *A History of the Warfare of Science with Theology in Christendom*, 2 vols, New York: Appleton.

Wilkins, John (1684) *A Discourse concerning a New Planet, Tending to Prove 'tis Probable our Earth Is One of the Planets*, London.

Woods, J. (1996) "Deep Disagreements and Public Demoralization," in D.V. Gabbay and H.J. Ohlbach (eds) *Practical Reasoning*: 650–62, Berlin: Springer.

——(2004) *The Death of Argument*, Dordrecht: Kluwer.

INDEX

Printed in the United States
by Baker & Taylor Publisher Services

Printed in the United States
by Baker & Taylor Publisher Services